Scientific Methods: Conceptual and Historical Problems

Edited by

Peter Achinstein
Laura J. Snyder

KRIEGER PUBLISHING COMPANY
MALABAR, FLORIDA
1994

Original Edition 1994

Printed and Published by
KRIEGER PUBLISHING COMPANY
KRIEGER DRIVE
MALABAR, FLORIDA 32950

Library of Congress Cataloging-In-Publication Data
Scientific methods : conceptual and historical problems / edited by Peter
Achinstein, Laura J. Snyder.—Original ed.
   p. cm.
  Includes index.
  ISBN 0-89464-822-5
    1. Science—Methodology—History.   2. Science—Philosophy—His-
tory.   3. Physics—Methodology—History.   4. Physics—Philosophy—
History. I. Achinstein, Peter.   II. Snyder, Laura J.
Q174.8.S36 1994
502.8—dc20   93-47265
                                                              CIP

10 9 8 7 6 5 4 3 2

CONTENTS

CONTRIBUTORS

Peter Achinstein, professor of philosophy at Johns Hopkins University, was the director of the 1992 N.E.H. summer seminar for college teachers on methodological debates in 19th century physics. His most recent book, *Particles and Waves* (New York, 1991), shared the 1993 Lakatos prize for an outstanding contribution to the philosophy of science.

Michael Bishop is assistant professor of philosophy at Iowa State University. He received his Ph.D. from University of California, San Diego, and has published papers in the philosophy of science.

Xiang Chen is assistant professor of philosophy at California Lutheran University. He received his Ph.D. from Virginia Polytechnic University, and has published papers in the history and philosophy of physics.

Dale Lynn Holt is assistant professor of philosophy at Mississippi State University. He received his Ph.D. from Vanderbilt, and has published in various philosophical journals.

Barbara Horan is assistant professor of philosophy at University of Maryland, Baltimore County. She received her Ph.D. from University of Minnesota, and has published various articles in the philosophy of biology. She is currently completing her book *Optimality Models in Evolutionary Biology.*

Niall Shanks is assistant professor of philosophy at East Tennessee State University. He received his Ph.D. in philosophy from the University of Alberta, Canada. He has published various articles on the foundations of physics.

Jonathan Smith is assistant professor of English at University of Michigan, Dearborn. He received his Ph.D. in English from Columbia, and is the author of *Fact and Feeling: Baconian Science and the 19th Century Literary Imagination* (Wisconsin, 1994).

Laura J. Snyder was the N.E.H. seminar assistant, and is complet-

ing her Ph.D. in philosophy at Johns Hopkins University. Her
dissertation is concerned with historical and contemporary views of
scientific evidence. She received a B.A. from Brandeis and an M.A.
from Johns Hopkins.

INTRODUCTION

In the summer of 1992 a group of scholars interested in the philosophy and history of science joined together at Johns Hopkins University for an eight-week seminar sponsored by the National Endowment for the Humanities and directed by Peter Achinstein. The topic of the seminar was "Methodological Debates in Nineteenth Century Physics." We focused on three historical episodes: the debate between wave and particle theorists over the nature of light, James Clerk Maxwell's development of the kinetic-molecular theory of gases, and J. J. Thomson's discovery of the electron. These episodes raise fundamental philosophical problems about scientific method, including the nature of reasoning, experimentation, evidence, theory-change, and controversy in science. They also raise questions about the role of history of science in clarifying and resolving these problems.

The contributors to this volume, all of whom were participants in the seminar, discuss these issues in the papers that follow. Barbara Horan shows how Isaac Newton's rules of scientific method can be used (and were by Newton) to argue for the existence of unobservable entities such as the universal force of gravity. Michael Bishop develops a new contextual model of meaning for scientific concepts, and shows how it can be employed in understanding the controversy between Newton and Robert Hooke about whether light consists of particles or waves. Xiang Chen discusses the role of certain skills in designing and replicating experiments; he then applies his insights to the nineteenth-century debate over the nature of light. Treating an issue raised by Thomas Kuhn in the early 1960s, Niall Shanks explores the question of how revolutions in science occur. He discusses a 20th century attempt to account for quantum phenomena in classical, nonquantum ways. Laura Snyder criticizes a standard, "historical" view of scientific evidence, which holds that whether some fact confirms an hypothesis depends upon the time at which the fact is known relative to the invention of the hypothesis; she argues that any view of scientific evidence must be nonhistorical.

Jonathan Smith demonstrates the influence of philosophical discussions of scientific method in areas outside of science, particularly in the work of the nineteenth-century art critic John Ruskin. Finally D. Lynn Holt, who discusses the relation between history and philosophy of science, argues that narrative history is necessary for establishing scientific claims.

The papers are intended to introduce these topics in the philosophy of science to students and other non-specialists; they should also be of interest to those working in the field.

The editors are grateful to Gordon Patterson of Krieger Publishing Company for suggesting this volume. Judy Kopec provided invaluable assistance in preparing the manuscript. We also wish to thank the NEH for sponsoring this seminar for college teachers, which allowed younger scholars time not only for research but also communal reflection, both in the seminar meetings and during many late nights which followed.

1

INFERENCE TO THE UNOBSERVABLE: NEWTON'S EXPERIMENTAL PHILOSOPHY

Barbara L. Horan

ABSTRACT

How can we establish the existence of unobservable entities if all of our evidence must come from what can be observed? Newton's experimental philosophy, structured by the four Rules of Reasoning presented in the Principia Mathematica, *offered a compelling solution to this problem. In this paper I explain Newton's view that belief in the existence of an unobservable entity is reasonable when that entity can be shown to be a* vera causa, *or true cause, that is, when its existence can be "deduced from the phenomena." I also discuss why he thought an alternative method, the method of hypothesis, could not justify belief in unobservable entities. I use Newton's derivation of the law of universal gravitation, which postulated the existence of an unobservable, universal force of attraction, to illustrate the use of his four Rules of Reasoning in deducing the existence of an unobservable entity from the phenomena.*

THE PROBLEM OF UNOBSERVABLE ENTITIES

The history of science is littered with discarded proposals of ethers, humors, spirits, and fluids. Scientists propose the existence of all kinds of things, only of some of which turn out to exist. Phlogiston, no; neutrinos, yes. Many of the things they propose are entities, processes or forces that, at the time they are introduced, cannot be observed. Some of these entities come to be observable as instrumentation evolves. Such was the case with molecules and electrons. Ethers and humors were never observed, and probably never will be. Of course, we can't look into the future to determine whether a scientific proposal of an unobservable entity will be borne out by eventual observation. We need a way of telling, when the proposal is first made, whether it is a reasonable one to believe.

For the purposes of this paper, I will take the scope of observability to be defined by contemporary scientific practice, assuming that in individual cases the goals and standards of observability are well understood, though perhaps not explicitly stated. Thus I will say that with the invention of the cathode ray tube and the scanning electron microscope, electrons and molecules became observable to practicing physicists and biologists. Entities that are postulated to exist but which are at the time beyond the limits of detection by the human senses aided by scientific instruments are unobservable entities. Under what conditions, if any, are we justified in believing in the existence of unobservable entities?

Newton had an answer to this problem. Actually, he had a whole philosophical system to answer this problem. He set down in the *Principia* some guidelines by which scientific reasoning was to proceed—his four "Rules of Reasoning in Philosophy" (in Newton's day science was called "natural philosophy"). The first of these rules stated the principle of *vera causa*, according to which "We are to admit no more causes of natural things than such as are both true and sufficient to explain the appearances" (1966, 398). Newton's solution to the problem of unobservable entities rested on this principle. He held that belief in the existence of an unobservable entity was reasonable just when that entity could be shown to be a *vera causa*.

In this paper I will first describe Newton's experimental method, and explain how his four Rules of Reasoning were to be used in establishing the existence of unobservables. I will then show how Newton applied this method to infer the existence of the unobservable force of universal gravitation. Finally, I will offer a brief assessment of Newton's method.

THE PRINCIPLE OF VERA CAUSA

Newton's principle of *vera causa* embodied two of the most important ideological and methodological changes that characterized the scientific revolution of the sixteenth and seventeenth centuries. These were (1) the new role for observation and experiment in justifying scientific claims, and (2) the concomitant rejection of *a priori* certainty as the ideal for scientific knowledge (Osler 1970).

For Galileo, experimentation had been mainly a rhetorical tool, a means of illustration and persuasion. His scientific conclusions were often reached by *a priori* deliberation, rather than by experimental inquiry. Descartes adopted the mathematical-deductive approach of Galileo and wielded it as a principle of scientific reasoning known as the "method of hypothesis." According to this method, a scientific proposition could be accepted as true so long as it explained a range of observed phenomena. Descartes recognized that in many cases it was possible to give more than one explanation of what was observed. Accordingly, he emphasized the need for experimentation, both to supply additional observations that might narrow the range of possible explanations, and to rule out candidate explanations by showing how the phenomenon might be absent even though the cause were present. But Descartes would have accepted any one of several mechanisms that conformed to the general "schemata of explanation" dictated by his theory. That is, he would have considered as explanatory any one of several hypotheses that "saved the phenomena," provided that they also satisfied general constraints imposed by his metaphysics, for example, that they have a geometrical formulation (Williams 1978, 264). Newton, by contrast, categorically rejected the method of hypothesis. He regarded experimental investigation as indispensable in proving explanatory principles. How were his experimental proofs made? And why did he insist on the *experimental* proof of scientific principles?

Newton's experimental proofs proceeded in two stages. In the first, inductive stage, one employed what was called the "method of analysis." Analysis consisted in inferring or "deducing" propositions from the phenomena, and then generalizing these propositions by induction. Newton was not explicit about what type of arguments would qualify as a "deduction from the phenomena," and his use of the term "deduction" to describe a diverse set of inferences, some of which have an inductive character, is confusing. Moreover, his only explicit definition of "phenomena" occurs in unpublished manuscripts.[1] There he describes phenomena as "whatever can be seen and is perceptible." In his published writings Newton clearly allows as phenomena both what can be directly observed and what is inferred from observations, but what seems more important than their direct or inferential relation to observation is their uncontroversial character. Phenomena are facts that everyone would agree upon, or that anyone could determine given that they took the trouble to make the appropriate observations.

For an understanding of the type of inferences Newton allowed as deductions from the phenomena, we must rely on examples from the *Principia* and from his later work, the *Opticks*. In an important article on Newton's experimental philosophy and the particle theory of light, Peter Achinstein (1991) describes three kinds of inferences to be found in Newton's discussions. First, there are ordinary deductive inferences, for example, the sort that would be found in mathematical proofs. For instance, Newton used the mathematical proposition that every body moving with uniform circular motion about a fixed point will be drawn to that point by a "centripetal" force, together with the fact (phenomenon) that Jupiter's moons move in such orbits, to infer that Jupiter's moons are subject to a centripetal force that keeps them in their orbits (1966, 42, 401, 406).[2] Second, there are inductive inferences. By "inductive inference" Newton meant an inference from the properties of observed members of a class to the properties of other, or of all, members of the class. Inductive inferences are prescribed by the third Rule of Reasoning: "The qualities of bodies, which admit neither intensification nor remission of degrees, and which are found to belong to all bodies within the reach of our experiments, are to be esteemed to be the universal qualities of all bodies whatsoever" (1966, 398).[3] As an example, he argued that all bodies must have inertia (a tendency to stay in motion, once in motion, and to stay at rest, once at rest) because all the bodies we have ever observed have inertia (1966, 398). Third, there are inferences Achinstein describes as "causal simplification." These are licensed by Newton's second Rule of Reasoning. This rule states that "to the same natural effects we must, as far as possible, assign the same causes" (1966, 398). When we observe the same or similar effects in different systems, Rule 2 tells us to simplify the causal picture and assign the same cause to the effects in each case. For instance, upon observing that the moons of Jupiter move in elliptical orbits, and observing that the moons of Saturn move in elliptical orbits, we are to infer that the cause of the motion in one case is the same as the cause of the motion in the other case. Similarly, Rule 2 requires that we identify causes that act in similar ways. If centripetal forces and gravitational forces obey similar laws and produce similar effects, Newton would argue, then they must be instances of the same kind of force.

In the second, deductive stage of experimental reasoning one applied the "method of synthesis." There is a striking similarity between the method of synthesis and the method of hypothesis, for in synthesis one showed that a proposition, if true, could explain a range of observed phenomena. However, the two methods are distinct. According to the method of synthesis, one *completed* the justification of a scientific proposition that had been deduced from the phenomena by showing that it could explain some set of observations. According to the method of hypothesis, the ability of a hypothesis to explain the observations was sufficient, by itself, for justification. No evidence independent of its explanatory success was required. Newton insisted that the general causal principles employed to explain the appearances of things must first be deduced from the phenomena: synthesis must always be preceded by analysis. Accordingly, he held that the method of hypothesis could not establish *verae causae*. Nor could it lead us to reject explanatory principles that had been deduced from the phenomena.

Newton's fourth Rule of Reasoning can be understood in this connection. "In experimental philosophy," the Rule states, "we are to look upon propositions inferred by general induction from phenomena as accurately or very nearly true, notwithstanding any contrary hypotheses that may be imagined, till such time as other phenomena occur, by which they may either be made more accurate, or liable to exceptions" (1966, 400). And Newton added: "This rule we must follow, that the argument of induction may not be evaded by hypotheses." Here again Newton's terminology is confusing. Although he used the term "induction" in stating Rule 4, his expression "induction from phenomena" should be understood as referring to the reasoning outlined by the method of analysis, that is, to what he otherwise calls "deduction from the phenomena." Under this interpretation, Rule 4 can be seen to play two roles in Newton's experimental philosophy. First, it asserts that experimental reasoning provides scientific claims with empirical support of the greatest possible strength. Conclusions reached by experimental reasoning are to be regarded as accurate or very nearly true. Second, it claims that this evidence cannot be weakened by competing hypotheses that might be imagined as alternative explanations of what has been observed. The conclusions of experimental reasoning cannot be evaded on the grounds that other explanations are possible.

Newton used his experimental method—analysis followed by synthesis—to prove the existence of unobservable entities. Rule 1 allowed him to infer the existence of an unobservable entity which could be shown to be a *vera causa*. To show that an entity was a *vera causa* was to show that it figured in a successful explanatory principle as a cause whose existence could be deduced from the phenomena. But how could Newton deduce the existence of *unobservable* causes from *observed* phenomena? He could do so by first

employing Rule 3, the rule of induction, to infer the existence of
properties of unobserved, and even unobservable entities. He could
then assume that these properties were effects produced by some
unknown, unobservable cause whose identity he could discover by
the causal simplification permitted by Rule 2. That is, once he knew
the cause of some set of observed phenomena, he could reason by
Rule 2 that the same type of cause must be responsible for similar
effects, *even if that cause and its effects were themselves unobservable.*
And by Rule 4 he could claim that the existence of such a cause was
very probable, regardless of the possibility that those effects could be
explained in some other way.

The inability of explanatory success to discriminate among various hypotheses proposed to account for observed phenomena created
pressing problems as the scientific revolution moved toward postulating the unobservable to explain the observable. For example, the
size, shape, and velocity of sub-microscopic—hence, unobservable—
material particles had been invoked by the chemist Robert Boyle to
explain the texture, color, and motion of bodies. But how were claims
about such unobservable properties to be justified? What rules
governed inferences to their existence? I am suggesting that Newton's principle of *vera causa* supplied an answer to these questions.
The properties of Boyle's sub-microscopic particles would not have
qualified as *verae causae* for Newton, because their existence could
not be shown to be logical consequences of previously established
principles, nor could they have been inferred by induction from what
could be observed, nor could their existence have been made certain
by any form of causal simplification. The only appropriate attitude
toward their existence would have been the tolerant agnosticism
prescribed for all hypotheses advanced without experimental proof.

Newton readily admitted that his method would not guarantee the
truth of any scientific proposition. Experimental reasoning "proved"
propositions only to the extent that it showed that they were very
probable, or likely to be true. It could not give them the certainty of
first principles known to be true *a priori.* Furthermore, not every
proposition admitted even of this qualified, empirical "proof." The
experimental method could prove propositions involving *verae causae.* The rest, including some of Newton's most cherished ideas,
remained hypotheses, plausible but nonetheless conjectures.

Newton was sometimes misunderstood on this point. In his own
experiments on color he had used a prism to separate sunlight into

rays of different colors, and then, collating them with a second prism, had reconstituted white light. The two principal ideas of his theory of color—first, that colors appeared in the decomposition of white light and not by the modification of light, as most others maintained, and second, that individual rays of colored light, whose properties were unchanged by reflection and refraction, differed from one another in their refrangibility (that is, in their refractivity)—he claimed to have inferred from experimental observations. They were not hypotheses. When Robert Hooke and Christiaan Huygens suggested that his theory was best regarded as a "probable hypothesis," Newton rankled. ". . . [M]y design was quite different," he wrote to Henry Oldenburg, President of the Royal Society, "and it seems to contain nothing else than certain properties of light which, now discovered, I think are not difficult to prove, and which if I did not know to be true, I should prefer to reject as vain and empty speculation, than acknowledge them as my hypothesis" (1959, 144).[4]

By contrast, his attitude toward the "corpuscular hypothesis," the central proposition of his particle theory of light, was more cautious. According to this hypothesis, rays of light consist of small bodies, or corpuscles, emitted from shining substances. Newton neither claimed to have deduced the corpuscular hypothesis from the phenomena, nor did he give it any great epistemological weight. In particular, he did not assign to light corpuscles the status of *verae causae*. In the same letter to Oldenburg, he complained:

> 'Tis true that from my theory I argue the corporeity of light, but I do it without any absolute positiveness, as the word *perhaps* intimates, and make it at most a very plausible consequence of the doctrine, and not a fundamental supposition. . . . Had I intended any such Hypothesis I should somewhere have explained it. But I knew that the properties which I declared of light were in some measure capable of being explicated not only by that, but by many other mechanical hypotheses. And therefore I chose to decline them all, and speak of light in general terms . . . without determining what that thing is. . . .(1959, 174)[5]

For Newton, all reasoning about causes, whether mechanical or occult, whether observable or unobservable, was on a par as long as those causes were unproved by experiment. He refused to allow that an explanatory proposition might be regarded as true or even likely to be true if its evidential support came only from its ability to "explain," that is, deductively entail, some set of observations. What were Newton's reasons for rejecting the method of hypothesis as a pattern of reasoning that could demonstrate the existence of *verae causae?* The one most widely discussed by historians and philosophers of science is suggested by his remarks that there will always be many competing hypotheses that can explain (in the sense that they deductively entail statements describing) a range of phenomena. Grimaldi's diffusion hypothesis, Hooke's undulatory hypothesis and Descartes' pression hypothesis, he observed, could all explain the phenomena created by his prism experiments.[6] None of these hypotheses, nor of course his own corpuscular hypothesis, could be regarded as true on this basis alone. Clearly, all of them satisfied the explanatory condition, but at most one of them (and perhaps none of them) was true. Therefore it could not be only the fact that a hypothesis satisfied the explanatory condition that made belief in it reasonable. Without prior analysis, that is, deduction from the phenomena, synthesis showed only what were *possible* causes, not what were the real causes.

There may be two other reasons why Newton rejected the method of hypothesis as a means for establishing *verae causae*. In a letter to Oldenburg written as a reply to Pardies' objections to his theory of color, Newton reiterated the point that his theory of color states the important properties of light without advancing any hypothesis. He then added:

> . . . the best and safest method of philosophizing seems to be, first to inquire diligently into the properties of things, and establishing those properties by experiments and then to proceed more slowly to hypotheses for the explanation of them. *For hypotheses to be explanations of the properties of things, they must only be fitted to those properties and not used to determine them, except insofar as they can furnish experiments.* (Cohen 1958, 106)[7]

This important passage points to a role Newton might have been willing to allow hypotheses to have in his experimental philosophy: hypotheses can sometimes suggest experiments. The supposition that light is a wave upon the ether suggested that the appropriate experiments to perform were those that might determine whether light had wave-like properties. For example, just as water waves bend around edges, and sound waves bend around obstacles, light, if it is a wave, should bend around corners and into the shadow created by objects on which it falls.

But this feature of hypotheses, that they can suggest the properties we can expect things to have and might detect by experiment, is also a danger. If we adopt the method of hypothesis, it is possible that our conjectures will determine the properties we observe them to have, either in the sense that we allow the hypothesis to pick out what are the relevant properties calling for explanation, or in the sense that the hypothesis actually influences our conception of what those properties are. An illustration of this danger was provided by Newton's contemporary, Robert Hooke, who was convinced that light was composed of waves. He could not understand why colors had to be already present in white light, rather than being modifications of light in the way that sounds were modifications of air:

> But why there is a necessity, that all these motions, or whatever else it be that makes colours, should be originally in the simple rays of light [as Newton claimed] I do not yet understand the necessity; no more than that all those sounds must be in the air of the bellows which are afterwards to heard to issue from the organ pipes, or in the string which are afterwards by differing stoppings and strikings produced... (1959, 111)[8]

The method of hypothesis will thus sometimes blind us to the real properties of things: a hypothesis may lead us to perceive as the properties things have just those properties it would have if the hypothesis were true.[9] Moreover, it can hardly be satisfying to have as an explanation a hypothesis that has itself determined which properties are to be explained. The fact that the wave hypothesis explains the wave-like properties of light is neither very surprising nor very interesting if the only reason we believe that light has wave-like properties is that we have adopted the wave hypothesis. Furthermore, if the method of hypothesis is allowed, then every hypothesis will be explanatory, for hypotheses will dictate what the properties of things are, and those properties will be precisely those for which the hypothesis can provide an explanation. This will not happen if we first look carefully at the phenomena, and then cautiously introduce explanatory hypotheses. For then hypotheses will have to be adapted to the properties we have independent reasons for believing things to have.

There may be a third reason for Newton's reluctance to accept the method of hypothesis. In another letter to Oldenburg, Newton once again pointed out that many hypotheses may be invented to explain the same set of observations. The mere possibility of the truth of a hypothesis cannot be a guide to true causes and real properties, nor to the empirical "certainty" of experimental inquiry. But he went on to give a different reason for rejecting the method of hypothesis.

Without the anchor of empirical proof, hypotheses are easily altered to accommodate falsifying observations:

> . . . if the possibility of hypotheses is to be the test of the truth and reality of things, I see not how certainty can be obtained in any science; *since numerous hypotheses may be devised, which shall seem to overcome new difficulties.* Hence it has been here thought necessary to lay aside all hypotheses, as foreign to the purpose, that the force of [Pardies'] objection should be abstractedly considered, and receive a more full and general answer. (Cohen 1958, 106, my italics)[10]

He repeats this idea in another passage:

> After the properties of light shall, by these and such like experiments, have been sufficiently explored . . . hypotheses are thence to be judged of, and those to be rejected which cannot be reconciled with the phenomena. *But it is an easy matter to accommodate hypotheses to this doctrine.* For if any one wish to defend the Cartesian hypothesis, he need only say that the globules are unequal, or that the pressures of some of the globules are stronger than others, and that hence they become differentially refrangible, and proper to excite the sensation of different colours. (Cohen 1958, 108, my italics)[11]

The method of hypothesis encourages what philosophers of science have called the "ad hoc rescue" of falsified hypotheses. When testing a hypothesis against observations, it is considered legitimate to revise the hypothesis if some of its predictions were incorrect. What is not considered legitimate is to revise the hypothesis for the sole purpose of accommodating those observations, and then to claim that the explanatory success of the hypothesis counts as evidence for its truth. Such revisions are considered arbitrary and "ad hoc." One wants theoretical revision to be well motivated, either by the goal of elaborating the theory's basic postulates or models, in order to reconcile the theory with other widely accepted theories, or because it will enable the theory to make predictions of new phenomena. Ad hoc modification of a theory falls outside this class of well-motivated changes. It is a revision of a theory carried out solely to allow the theory to avoid refutation by a false prediction. According to the method of hypothesis, however, all methods of proposing hypotheses are on a par. In fact, hypotheses that are revised in order to explain a greater number of observations are preferred. On the method of hypothesis, therefore, ad hoc modification of falsified hypotheses is a virtue, not a flaw!

THE EXISTENCE OF UNIVERSAL GRAVITATION

Universal gravitation, the force of attraction that exists between bodies separated by empty space and that is diminished, but never entirely gone, across great distances, could not be observed either by the unaided senses or by the scientific instruments of Newton's day. The familiar force of terrestrial gravity was widely believed to be a property possessed by falling bodies, but it was not regarded as a universal property of matter, in the sense that all bodies possessed it. Heavy bodies fell because they possessed gravity, but lighter bodies, like air and smoke, rose because they possessed "levity." Nor was gravity regarded as a single, unique property. In antiquity gravity had been thought to be a sort of natural affinity or kinship of like pieces of matter. Similarly, natural philosophers of the Renaissance, like Copernicus, Galileo, Kepler and Gilbert, regarded gravity as a property peculiar to each type of matter. Pieces of Mars would tend towards other pieces of Mars, and pieces of Earth would tend toward pieces of Earth, but no one supposed that pieces of Mars might fall towards pieces of Earth. Various ideas about how gravity worked in bodies had been proposed. Descartes held that gravity was created by an ether vortex turned by the Earth. His contemporary, Pierre Gassendi, thought that bodies were connected to the Earth by thin strings; large bodies weighed more than small bodies because they were connected by more strings (Gjertsen 1986, 236–37).

Although terrestrial gravity was a familiar force, a force of "celestial" gravitation was not. Kepler had suggested that some sort of magnetism exerted by the sun kept the planets in their orbits. But to many others the idea that gravity should be an attractive force that acted between bodies separated by empty space was absurd and even occult (*ibid.*). No one imagined that there was any connection between the property that caused heavy bodies to fall to the Earth (or other planet) and whatever it was that kept planets in their orbits. Certainly no one had observed what Newton would prove to exist: a single, universal force that caused the motion of falling bodies, kept objects from flying off the surface of the Earth, caused the planets to remain in their orbits around the sun, caused satellites to remain in orbit around planets, caused comets to make periodic returns to the solar system, and caused the tides (Cohen 1985, 164). How did Newton establish the existence of this force?

One of the great triumphs of Newton's *Principia* was the explanation of the motion of the planets, the comets, the moon, and the sea by the law of universal gravitation, a law that he regarded as an explanatory principle deduced from the phenomena of terrestrial and celestial motion. In Book I of the *Principia* he introduced three important laws of motion. According to the Law of Inertia, all bodies possess inertia, or resistance to movement. Bodies at rest stay at rest, and bodies in rectilinear (straight-line) motion remain in motion, unless they are acted on by an outside force. An accleration is a departure from inertial motion, and requires an outside (non-inertial) force. According to the Force Law, the force required to accelerate a body is proportional to the mass of the body. Large bodies require more force to move than small ones. Newton showed for any given body, the ratio of force applied to acceleration created, F/a, is a constant, which he called the mass m of the body. (Thus the second law is often stated as $F = ma$). The Law of Action and Reaction states that for every force exerted by one body on another, there will be an equal and opposite force exerted by the second body on the first. "If you press a stone with your finger," Newton wrote, "the finger is also pressed by the stone" (1966, 13–14). He claimed that these three laws of motion had been deduced from the phenomena. From these laws and some well-established celestial phenomena, Newton deduced the law of universal gravitation. His argument, contained in Books I and III of the *Principia,* is elegant.[12]

According to the Law of Inertia, inertial motion is the natural condition of bodies, including planets. But clearly planets do not move in straight lines—they move along elliptical paths. Their motions may therefore be thought of as a combination of two motions, one inertial or rectilinear, and the other at right angles to the first, drawing the planets into their orbits. This second component of planetary motion, since it is not inertial motion, must be produced by a force. Newton's task was to calculate the magnitude of this force.

His first step was the proof that for a body moving with purely inertial motion, a line connecting the body to a fixed point will sweep out equal areas in equal times. He proved that if to a body's inertial motion was added a "centripetal" force (a forced directed from the body toward that point), the line connecting that body to the point would also sweep out equal areas in equal times. He then demonstrated the important converse: if the radius of a body's orbit swept out equal areas in equal times, then it must be acted upon by a centripetal force.

According to Kepler's first law of planetary motion, the planets moved in elliptical orbits around the sun. According to Kepler's second law, each planet's orbital radius swept out equal areas in equal times. Newton knew from the work of other astronomers that the moons of Jupiter and Saturn orbited those planets in a similar fashion, that is, in elliptical orbits with radii sweeping out equal areas in equal times. Applying the mathematical propositions he had proved to these phenomena, he concluded that there should be centripetal forces from the sun to each planet, and from Jupiter and Saturn to their moons. Having deduced these forces from the phenomena, he could infer that they were *verae causae,* or causes with real, not merely hypothetical, existence.

To establish the magnitude of these forces Newton again looked to the observations of astronomers for the phenomena. Kepler's third law stated that the orbits of the planets around the sun were such that the ratio r^3/T^2 was a constant (r is the radius of each planet's orbit; T is the period of the orbit, or the time required for one complete revolution around the sun). He knew from various reports that the moons of Jupiter orbited the planet in such a way that they too obeyed Kepler's third law. Newton then proved that in any system of bodies obeying Kepler's third law each orbiting body would be subject to a centripetal force whose strength was inversely proportional to the square of its radius, that is, $F \propto 1/r^2$. His proof rested on the three laws of motion, geometrical considerations, and mathematical assumptions about the behavior of vanishingly small quantities. With the aid of this proposition he derived the result that planetary satellites, as well as planets themselves in orbit around the sun, were acted on by centripetal forces whose magnitude varied as the inverse square of their distance from the center of their orbits.

What were these centripetal forces? In his book the *Opticks,* published late in his life, Newton reported that in 1666 he had begun to reflect on the possibility that the force of terrestrial gravity extended as far as the moon. Could it be one and the same force that caused bodies to fall and caused the inertial motion of the planets to be deflected into elliptical orbits around the sun? He may have had this thought while sitting in his garden in Lincolnshire, where he had gone to escape the plague that by that time had spread to Cambridge. This is the setting, at least, for the story that the idea of universal gravitation came to him after seeing the fall of an apple (Cajori 1962, 64). Twenty years later, in the *Principia,* he published his proof that terrestrial gravity and the centripetal force of planetary orbits were identical.

Newton used Kepler's third law to determine that the centripetal force drawing planets in towards the sun was "pretty nearly" the same strength as the force of gravity at the surface of the earth. This was in 1666. To make the case that terrestrial gravity was the same force as the one that controlled planetary orbits, of course, he needed exact agreement. This he accomplished by proving that the centripetal force exerted by a large, homogeneous sphere on a small object—a calculation he could not complete—would be equal to the force exerted by a large body all of whose mass was concentrated at its geometric center. This calculation he could make. It is reported in Book III of the *Principia*.

From observations of the motion of the Earth's moon, Newton knew that the radius of the moon's orbit swept out equal areas in equal times. The moon's orbit must therefore be maintained by a centripetal force directed towards the earth. The earth has only one moon; however, Newton reasoned by Rule 3, the rule of induction, that if it had several moons, the ratio r^3/T^2 would be constant. That is, the Earth's moons, like the moons of Jupiter and Saturn, would satisfy Kepler's third law. From this result, together with the proposition that any orbiting body satisfying Kepler's third law must experience a centripetal force whose magnitude is proportional to inverse square of the orbital radius, Newton concluded that the centripetal force of the earth acting on the moon must be proportional to $1/r^2$.

But what of gravity? Ptolemy, Copernicus, and Tycho Brahe had all made estimates of the distance of the moon from the Earth. Precise estimates of the period of the moon's orbit were known. With this information Newton computed the distance the moon would "fall" towards the Earth if released from its orbit. When close to the Earth, the distance it would fall in a short time as a result of its centripetal force turned out to be *exactly* equal to the distance ordinary bodies near the earth's surface would fall in the same time.[13] Newton's second Rule of Reasoning required that two causes acting in similar ways and producing similar effects be regarded as instances of the same type of cause. Gravity was the cause of bodies falling near the Earth's surface. The centripetal force acting on the moon would produce exactly the same effect were the moon close enough to the Earth. This centripetal force therefore must *be* gravity!

> Therefore since both these forces, that is, the gravity of heavy bodies, and the centripetal forces of the moons, are directed to the center of the earth, and are similar and equal between themselves, they will (by Rule 1 and 2) have one and the same cause. And therefore the force which retains the moon in its orbit is that very force which we commonly call gravity. . . .
> (1966, 409)

Newton immediately turned to the orbits of the planets around the sun and to the orbits of moons around the planets. All of them manifested appearances similar to the orbit of the moon around the earth, for example, they obey Kepler's laws. The moon's orbit is controlled by gravity and by Rule 2 he could infer that the centripetal forces that kept the planets in their orbits around the sun, and that kept moons in their orbits around the planets, must also be gravitational forces.

Furthermore, according to the Law of Action and Reaction, all attraction is mutual. The planets gravitate toward the sun; the sun must therefore gravitate toward the planets. Newton then observed that we know from experience that bodies fall at the same rate, regardless of their weight. By Rule 3 he inferred that this must be true of all bodies. In addition, the Force Law implies that when unequal masses are equally accelerated, unequal forces must be acting. The gravitational attraction between bodies must therefore vary according to their masses. The final result was Newton's law of universal gravitation, an explanatory principle that invoked an unobservable force of attraction existing between all bodies in the universe whose magnitude varied as the product of their masses and inversely with the square of the distance between them.

CONCLUSION

I have argued that Newton's experimental philosophy offered an interesting solution to the problem of how to reason from the observable to the existence of the unobservable. His views on scientific method dominated thinking in many fields well into the nineteenth century. However, his method of experimental proof was not universally accepted. At the time his work was published, Newton was engaged in a heated debate about the nature of light. As we have seen, he defended the view that light was composed of particles, while his opponents, the wave theorists, defended the view that light was composed of waves. By the middle of the nineteenth century wave theorists had offered plausible explanations of phenomena like partial reflection, refraction, diffraction and interference effects, some of which either could not be explained at all by the particle theory, or could be explained only by making implausible, improbable auxiliary assumptions. The relatively poor performance of the particle theory may have led some people to doubt the effectiveness of Newton's method. Similar doubts may have been created when scientists finally recognized that Newton's theory of universal gravitation could not explain the anomalous behavior of the orbit of Mars. Although Newton's theory had received striking confirmation when it successfully predicted the existence of an unobserved planet (Neptune) as the cause for the perturbations of Uranus' orbit, the existence of the unobserved planet Vulcan as the cause of anomalies in Mars' orbit was never verified. By the late nineteenth century, it was clear that Newton's theory could not explain the observed motions of the planets, and this conclusion naturally cast suspicion on the methodology that had led to the development of that theory.

Moreover, in the period from roughly 1830 to 1860, the virtues of Newton's experimental method were explicitly debated by defenders of the method of hypothesis. They raised serious objections to Newton's assertions that the pattern of inductive and deductive reasoning he described could establish the causes of observed phenomena. For example, William Whewell, the Cambridge polymath, historian, and philosopher, argued that inductive methods would never permit inference to the existence of new causes, since deductions from the phenomena always proceeded from what had already been observed. And he defended the method of hypothesis as a means of establishing explanatory principles, even those referring to unobservables.

The soundness of Newton's method of experimental reasoning was thus challenged both by the apparent failure of his physical and optical theories, and by direct criticism from proponents of a rival methodology. However, none of Newton's critics succeeded in meeting the objections he raised to the method of hypothesis—its inability to discriminate true causes from imaginative fictions, its justification of hypotheses that only explained because they were allowed to fashion the properties to be explained, and its tolerance of ad hoc modification of falsified theories. The claim that the explanation of a range of observed phenomena is sufficient to make a hypothesis highly probable has never been convincingly argued. A thorough defense of Newton's experimental method is beyond the scope of this paper. But it may be said here that in view of the powerful inferences sanctioned by his four Rules of Reasoning, the unanswered objections to the method of hypothesis, and in the absence of any proof that this alternative method can succeed, Newton's experimental philosophy still remains the most promising solution to the problem of unobservable entities.

NOTES

1. See, for instance, the unpublished Rule V, discussed by I. B. Cohen (1978) in his *Introduction to Newton's Principia*, p. 30. A useful discussion of Newton's view of phenomena can be found in Peter Achinstein (1991), 33–35.

2. Newton, *Principia Mathematica,* Book I, Proposition II; Book III, Phenomenon I; Book III, Proposition I.

3. The restriction to qualities "which admit neither intensification nor remission of degrees" is difficult to understand. Some light may be shed on it by Newton's remark in Query 29 of the *Opticks* about magnetism and the special refracting properties of Iceland spar: "And as magnetism may be intended and remitted, and is found only in the magnet and in iron, so this virtue of refracting the perpendicular rays is greater in island crystal, less in crystal of the rock, and is not yet found in other bodies" (1979, 373–74). This suggests that qualities that admit of intensification or remission of derees—qualities to which inductive generalization may not be applied—are qualities like those of Iceland spar, that are found in varying degrees or, like magnetism, that are only found in certain kinds of bodies.

4. Newton to Oldenburg (Reply to Pardies' First Letter), 13 April 1672.

5. Newton to Oldenburg (Reply to Hooke), 11 June 1672. I have modernized Newton's spelling and capitalization.

6. Grimaldi discovered that when a beam of light passed through a small hole, the beam was diffused or "diffracted," a phenomenon he explained by supposing light to be a substance in very rapid motion. Hooke believed that vibrations of luminous bodies excited the aether and produced waves that spread and expanded as the light diffused. Descartes held that light was caused by a quick and violent movement in bodies whose pressure or "pression" was transmitted to our eyes in the same way that the resistance of bodies is transmitted to a blind man's hand through his cane (Gjertsen, 1986).

7. Newton to Oldenburg (Reply to Pardies' Second Letter), 10 June 1672. English translation (1672) published in *Philosophical Transactions,* Vol. VII, No. 85, 5014. Reprinted in Cohen (1958), 106–7. I have used my own translation for the italicized phrase *"Nam Hypotheses ad explicandas rerum proprietates tantum accommodari debent, et non ad determinandas usurpari, nisi quatenus experimenta subministrare possint."*

8. Letter from Hooke to Oldenburg (Critique of Newton), 15 February 1672.

9. Larry Laudan (1981, 90–91) discusses a similar objection made to the method of hypotheses by Thomas Reid.

10. Here I have followed the translation given by I. B. Cohen of Newton's Latin *"siquidem alias atque alias Hypotheses semper liceat excogitare, quae novas difficultates suppeditare videbuntur."* Other translators, for example, Turnbull, translate this passage as "if indeed it be permissible to think up more and more hypotheses, which will be seen to raise new difficulties." Cajori agrees, writing "it is always possible to contrive hypotheses, one after another, which are found rich in new tribulations" ("Appendix" to *Principia,* 673). I have chosen Cohen's translation because it allows the richest interpretation of Newton's rejection of the method of hypothesis.

11. Newton to Oldenburg (Reply to Pardies' Second Letter), 10 June 1672.

12. For an accessible modern treatment of Newton's derivation, see Cohen 1985.

13. Cohen (1985, 171) gives the agreement of the figures to within 0.0003 inch.

REFERENCES

Achinstein, Peter. 1991. "Newton's Corpuscular Query and Experimental Philosophy." In *Particles and Waves* (New York: Oxford University Press), 31–67.

Cajori, Florian. 1962. *A History of Physics,* revised ed. New York: Dover.

Cohen, I. B. 1985. *The Birth of a New Physics,* revised ed. New York: Norton.

Cohen, I. B. 1978. *Introduction to Newton's Principia.* Cambridge: Harvard University Press, 1–123.

Cohen, I. B., ed. 1958. *Isaac Newton's Papers and Letters on Natural Philosophy.* Cambridge: Harvard University Press.

Gjertsen, Derek. 1986. *The Newton Handbook.* New York: Routledge & Kegan Paul.

Laudan, Larry. 1981. *Science and Hypothesis: Historical Essays on Scientific Methodology.* Boston: Reidel.

Newton, Issac. 1959 [1661–1675]. *Correspondence of Isaac Newton,* Vol. I. Ed. H. W. Turnbull. Cambridge: Cambridge University Press.

Newton, Isaac. 1966 [1686]. *Mathematical Principles of Natural Philosophy and System of the World,* 2 vols. Trans. by A. Motte. Revised by F. Cajori. Berkeley: University of California Press.
Newton, Isaac. 1979 [1730]. *Opticks,* 4th ed. New York: Dover.
Osler, Margaret. 1970. "Locke and the Changing Ideal of Scientific Knowledge," *Journal of the History of Ideas* 31:3–16. Reprinted in J. Yolton (ed.), *Philosophy, Religion and Science in the Seventeenth and Eighteenth Centuries* (Rochester: University of Rochester Press, 1990).
Williams, Bernard. 1978. *Descartes: The Project of Pure Enquiry.* Atlantic Highlands, NJ: Humanities Press.

2
CONCEPTUAL CHANGE IN SCIENCE: THE NEWTON-HOOKE CONTROVERSY
Michael Bishop

ABSTRACT

We can decide between two leading models of scientific concepts by assessing how well they allow us to understand a real scientific controversy—the 1672 Newton-Hooke dispute about optics. According to semantic holism, *Newton and Hooke expressed concepts that presupposed their theories' controversial assumptions. Since any argument for one of those assumptions must have been circular, holism implies that scientific change isn't rational. The* cluster view *denies that all terms have singly necessary and jointly sufficient conditions of application. Some cluster views imply that a term might have different meanings in different contexts. So in one context, Newton might have expressed a concept that made controversial assumptions, and in another context, he might have expressed a concept that did not. The cluster model leaves open the possibility (without guaranteeing) that scientific change is rational. The context-sensitive cluster model provides a better understanding of the Newton-Hooke controversy than does holism.*

Philosophers worry about the nature of conceptual change in science because it is intimately connected to questions of truth and rationality. This paper will try to decide between two such views— semantic holism and the cluster view. If holism is true, then theory change in science cannot be rational. The cluster view, on the other hand, allows for the possibility (but does not guarantee) that scientific theory change is rational. I will argue that we should favor the cluster view over holism because it makes much better sense of actual scientific controversies, in particular, the 1672 controversy between Newton and Hooke about the nature of light and colors.

The first section introduces the holistic and cluster models and applies them to the Newton-Hooke controversy. The second section spells out some of the more radical implications of the holistic model. The next section sketches the optics necessary to understand the dispute. The fourth section hits the high points of the controversy, and argues that at each stage, the cluster model allows us to better interpret the arguments of both participants. The fifth section critically assesses arguments proposed by various Newton scholars in favor of the holistic reading of the conflict.

THE HOLISTIC MODEL VERSUS THE CLUSTER MODEL

Early in 1672, Isaac Newton published his first paper and immediately became embroiled in an unpleasant controversy with Robert Hooke. At issue was the nature of light and colors. Hooke had published his theory in 1665, and although it had some original features, it fit squarely in the dominant seventeenth-century optical tradition. This tradition, which included the theories of Descartes and Hobbes, was defined by two hypotheses. First, light consists of a wave-like motion that propagates through a pervasive medium, the ether; and second, white light is homogeneous, it is made out of the same basic kind of stuff, while colored light consists of modifications of white light. Newton's first article opposed both elements of the traditional view. Newton argued that light consists of moving particles. He also argued that each different color of light is homogeneous and that white light consists of a mixture of light of all different colors.

By the early part of the eighteenth century, Newton's theory had won the day. How did this happen? What factors led to the ascension of Newton's theory of light and colors? At least part of the common-sense answer is that Newton proposed evidence-based arguments for his theory that were *rationally compelling* at the time. Of course, other factors, including nonrational social factors, may have played a role. But the commonsense view of science is that the primary reason a theory becomes widely accepted is that there are good reasons to believe it.

The commonsense view has been challenged on a number of fronts. One important challenge comes from defenders of the holistic model of scientific concepts. In order to spell out semantic holism and its competitor (the cluster model), we need to reflect upon the nature of the meaning of scientific terms.

Meaning and Reference

In order to understand any view of conceptual change, we must distinguish between the *meaning* of a term on a particular occasion, and what that term *refers to* on that occasion. Suppose I point to an animal and say, "There is a doe." The meaning of 'doe' is the concept it expresses. The meaning of nouns can often be represented by a set of descriptions; for example, in this case, 'doe' might mean *a female deer*. What the term refers to in this case would be the animal. Notice that there are many properties possessed by the referent of a term that are not part of the term's meaning. The animal I pointed to might be especially fat, but this is not part of the meaning of 'doe'; in fact, the animal might have been a raccoon, in which case my original statement would have been false. So just because scientists use a term to *refer* to different things on different occasions, it does not follow that the *concepts* (or meanings) expressed by those terms are different.

Let's think about the nature of scientific concepts. When we learn a new scientific theory, our expressive capacity increases—we become able to understand and express brand new concepts. For example, after we learn the theory of relativity, we can employ our old terms (like 'time' and 'mass' and 'velocity') to express novel meanings. How is this possible? How can we attach brand new meanings to old words? And how is it possible for us to coin new words with new meanings (for example, 'spacetime')? The answer seems obvious enough: The meaning of a scientific term depends to a certain extent upon the theory in which that term finds itself.

Both models of conceptual change we will be considering accept this intuition. In order to fix the meaning of 'light ray' in Newton's idiolect, let's consider some of the statements Newton believed true of light rays.

1. *Irreducibility:* A light ray is the irreducibly smallest section of propagating light.
2. *Rectilinearity:* Light rays are rectilinear (that is, travel in straight lines) in the absence of obstacles.
3. *Finite velocity:* Light rays travel at finite velocities.

4. *Separability:* An obstacle can (in principle) separate a ray of light from other collateral and successive rays of light.
5. *Visibility:* Light rays are visible under normal conditions.
6. *Homogeneity:* Light rays of particular (nonwhite) colors are homogeneous.
7. *Refrangibility:* Light rays of different colors have different degrees of refrangibility.
8. *Heterogeneity:* White light consists of light rays of different colors.
9. *Particulate:* Light rays consist of particles.

For both the cluster and the holist models, the meaning of Newton's term 'light ray' will be given by some of these nine statements (and perhaps others I have overlooked).

The Holist View

According to the semantic holist, scientists express concepts that presuppose the controversial assumptions of their own theories. For example, the holist would argue that Newton and Hooke expressed concepts of light ray that presupposed their own theories of light and colors. So part of the *meaning* of Newton's term 'light ray' is that light is a particle and that white light is heterogeneous. And part of the *meaning* of Hooke's term 'light ray' is that light is a wave-phenomenon and that white light is homogeneous.

We can distinguish two ways a holist might fix the meaning of 'light ray' in Newton's theory.

1. *Strong semantic holism.* The meaning of 'light ray' is always given by the conjunction of all the statements in Newton's theory in which it occurs, 1–9.
2. *Weak semantic holism.* The meaning of 'light ray' always includes a subset of 1–9, say, the conjunction of 6–9.

Strong semantic holism is a considerably more radical view than weak semantic holism. The strong view implies that if Newton were to alter any one of the above nine statements, no matter how minor the change, the meaning 'light ray' would change as well. While there may be important differences between strong and weak semantic holism, I will ignore them for the purposes of this paper.

The holistic model of the Newton-Hooke controversy will hold that the meaning of 'light ray' for Newton was fixed by (at least) the homogeneity, refrangibility, heterogeneity and particulate assumptions. Whether its meaning is also fixed by all or some of the other assumptions is irrelevant for our purposes. The main point for the holist is that given the meaning of 'light ray' in Newton's idiolect, each of these four assumptions is necessary for something to be a light ray (if something lacks one of them, it can't be a light ray) and together they are sufficient (if something has all of those properties, it is a light ray). Central elements of this holistic view have been defended by Ronchi (1970), Sabra (1967), Shapiro (1975) and Buchwald (1989). No one has defended the complete holistic picture of this controversy.[1]

The Cluster View

The roots of the cluster view can be found in Wittgenstein (1953) and Searle (1958). At its core is the denial of the holist's contention that all terms have singly necessary and jointly sufficient conditions. To take Wittgenstein's famous example, the term 'game' does not have any necessary conditions since there is no single property that all games must possess. Further, the cluster view suggests that there will often be more than one set of jointly sufficient conditions for the proper application of a term. For example, tennis, catch, and chess are games, and yet they possess quite different sets of properties (Wittgenstein 1953, 66). Applied to Newton's expression 'light ray', the cluster view suggests that no single assumption (1–9) is necessary for something's being a light ray, and a number of different sets of those assumptions might be sufficient.

According to the cluster view, the defining properties of a term will not always be necessary for the proper application of that term. Therefore, the cluster view needs to specify a relationship between a term and its defining properties that is weaker than logical necessity. Peter Achinstein (1968) has articulated such a relationship. The defining properties (or assumptions) given in 1–9, above, are *semantically relevant* for light rays. We can understand semantic relevance roughly as follows. Suppose we want to know whether X is a light ray, and we already know that X has some of the properties (not including rectilinearity) on the above list. If we find out that X is rectilinear, this would *by itself* tend to count to some extent in favor of classifying X as a light ray. But if we discovered that X is not rectilinear, this would *by itself* tend to count to some extent against classifying X as a light ray. On the cluster view, the defining properties of some scientific terms will typically be semantically relevant (rather than necessary) for proper categorization.

The cluster view I want to defend differs from some other cluster views by being *context-sensitive:* A term might have different meanings on different occasions of use. In other words, in one context, Newton might have expressed a concept that assumes $S_1 - S_4$, and in a different context, he might have expressed a concept that assumes $S_5 - S_9$. As a matter of fact, however, I will argue (in the fourth section) that *the term 'light ray' as used by both Newton and Hooke can usually be understood in terms of the irreducibility, rectilinearity, finite velocity, and separability assumptions.*

CONSEQUENCES OF HOLISM

If we accept the holistic model for any particular controversy, at least three strange consequences ensue.

Proponents of different and competing theories use the same words with very different meanings. As a result, they will not be able to say the same thing about the world nor will they be able to contradict one another. For example, suppose we set up an experiment and Newton predicts, "The light rays will not bend into the shadow," and Hooke predicts, "The light rays will bend into the shadow." Although these appear to be contradictory, they really aren't because 'light rays' mean something different for Newton and Hooke. They are no more contradictory than "The bank [financial institution] is entirely red" and "The bank [riverside] is not entirely red." So what makes any experiment a setback for one theory and a victory for the other has nothing to do with their making contradictory predictions about the world, because such predictions are never made.

Theoretical statements are analytic. A statement is analytic just in case its truth depends solely on the meanings of its words. The typical example is 'Bachelors are unmarried.' A statement is synthetic just in case its truth depends both on the meanings of its words and the way the world is. 'Clinton is president' would be an example of a synthetic statement. If the meaning of scientific terms is fixed by the theories in which they are embedded, then theoretical statements become analytic. Newton's claim 'Light rays consist of particles' is true solely by virtue of the meaning of the term 'light rays' (as used by Newton). Further, in Hooke's mouth, the statement 'Light rays consist of particles' is false *by definition.*[2]

The arguments scientists make in favor of their theories will inevitably be circular. Newton argued for the conclusion that light consists of particles on the basis of the behavior of light rays. But if the holistic model is right, whenever Newton used the term 'light ray,' he was assuming that light is particulate. Newton's premises would presuppose exactly what is at issue. His argument would have been circular, and circular arguments are illegitimate. You can "prove" any proposition, no matter how absurd, if you are allowed to smuggle it into your premises.

Semantic holism has been defended by philosophers such as Paul Feyerabend (1975), Thomas Kuhn (1970) and Paul Churchland (1979). Some are likely to reject holism out of hand because of the above consequences. And they must seem odd to the defender of the commonsense view of science. Blatantly circular arguments will not be accepted by anyone who is not already committed to the truth of their conclusion; in fact, circular arguments should not be accepted by any rational agent. If the choice between different and competing scientific theories is always made on the basis of circular arguments, then it would seem that the rationality and objectivity of scientific progress go out the window. The answer to our original question—What factors led to the ascension of Newton's theory of light and colors?—cannot be that Newton offered rationally compelling arguments.

The holist will insist that this consequence is counterintuitive only if one is committed to a view of science that is overly optimistic about its essential rationality. Once we reject scientific utopianism, we will recognize the above consequences as obvious. If we are to reject semantic holism, we must do more than merely show that it has consequences which we find weird (and which holists willingly embrace). We must show in detail what is wrong with holism. I propose to do this by analyzing the 1672 Newton-Hooke controversy.

SCIENTIFIC BACKGROUND TO THE NEWTON-HOOKE CONTROVERSY

This paper will highlight two issues on which Newton and Hooke clashed: their physical models of light and their explanations of chromatic dispersion. In this section, I will sketch the scientific background necessary to understand and analyze the dispute.

Light bends when it passes from one transparent medium to another. This is called *refraction*. It explains, for example, why a

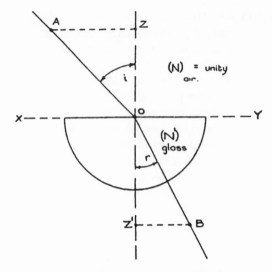

Figure 2.1 From Johnson 1960, 10. (Reprinted with the permission of Dover Publications, Inc.)

stick looks bent when placed in water. Eyeglasses (and contact lenses) also work by refraction; they bend light so that a clear image can be formed on the back inside part of your eyeball (the retina). Let's suppose that a ray of light (AO) traveling through the air enters a block of glass whose surface is given by XY (Figure 2.1). The angle i is called the angle of incidence and the angle r is called the angle of refraction. Is there some relationship between the angles of incidence and refraction? Well before Newton's day, natural philosophers knew that for any angle of incidence, the ratio between the sines of the angles of incidence and refraction is constant. (This is known alternatively as the law of refraction or the sine law or Snell's law.) As we shall see, the sine law played an important role in the Newton-Hooke controversy. It was well known that changing the media (the glass or the air) through which light travels changes the degree to which light is bent. (For example, given a specific angle of incidence, a light ray will be bent more by a diamond than by ice.) So it is important when testing the sine law to keep the media constant. The view Newton argued for in his first publication implied that changing the color of the light would also change the degree to which light was bent. So when testing the sine law, it is important not only to keep the media constant, but also to keep constant the color of the light. As we shall see, Hooke refused to accept the hypothesis that different colored light has different degrees of "refrangibility."

The behavior of light passing through prisms also played an important role in this controversy. Prisms in the seventeenth century were made of glass (as are many today). It was known that light will refract (bend) upon entering the prism (when passing from air into glass) and upon exiting the prism (when passing from glass into air); however, prisms have a more obvious influence on light. When a beam of white light passes through an appropriately placed prism, it fans out into a spectrum. This is known as *chromatic dispersion*. Rainbows are the most dramatic effect of chromatic dispersion. During a rain shower or when watering your lawn, water drops can act like prisms, thus separating sunlight into rainbows.

THE NEWTON-HOOKE CONTROVERSY

Now let's turn to the controversy between Newton and Hooke about the nature of light and colors. First, we will sketch Hooke's view of light and colors as described in his *Micrographia* (1665). Next, we will focus primarily on Newton's crucial experiment as described in his first publication (1672a); the crucial experiment was supposed to show that white light consists of rays of different degrees of "refrangibility." We will then touch on the high points of Hooke's (1672a) response to Newton, which was designed to show that Newton's experimental results could be explained by a number of different theories of light and colors. And finally, we will outline Newton's (1672b) angry rebuttal. I will argue that the cluster model does a much better job of making sense of this controversy than the holistic model.

Hooke's *Micrographia*

Hooke believed that light is a pulse propagated through an ether. This ether consists of tiny particles and is spread out everywhere that light can travel. Light is produced when a part of a luminous object vibrates and sends out pulses through the ether. In a homogeneous medium, every pulse generates a spherical wave front: ". . . necessarily every pulse or vibration of the luminous body will generate a sphere, which will continually increase . . ." (1665, 57).

Hooke, along with everyone else before Newton, believed that white light is basic, it is made out of a single kind of stuff, and colored

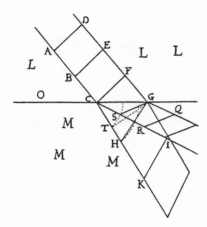

Figure 2.2. From Hooke 1665, Schem. VI., Fig.1. (Reprinted with the permission of Science Heritage, Ltd.)

light is some sort of modification of white light. Hooke's explanation of color begins with Figure 2.2. "We will suppose . . . ACFD to be a physical ray, or ABC and DEF to be two mathematical rays, trajected from a very remote point of a luminous body through an homogeneous transparent medium LLL" (1665, 57). For Hooke, a physical light ray is a line of light having width along which motion in the ether is propagated. A mathematical light ray is a line that is perpendicular to the tangent of a point on the wave front. Pulses of light (DA, EB, FC) are "small portions of the orbicular impulses which must therefore cut the rays at right angles" (1665, 57). Notice that in Figure 2.2, the pulses of light (AD, BE, CF) are perpendicular to the parallel light rays (AC and DF). While the ray that enters medium MMM (AC) travels from C to H, the other ray travels only from F to G. This causes the post-refraction pulses (GH and IK) to be oblique (not perpendicular) to the rays. According to Hooke's obliquity theory of color, whenever the pulse is oblique to the rays, colors are produced (blue along ray CK and red along ray GI). So blue and red light are modifications of white light, and all the other colors are produced by some mixture of blue and red light.

There are two factors that cause the prism's spectrum to disperse for Hooke. The first is diffusion. When physical ray ACFD first hits the new medium (MMM), the part of the pulse that hits the new medium first (the lower section of CF) is more "impeded by the resistance of the transparent medium, than the other part" of the pulse (the upper section of CF). The later part of the pulse "will be

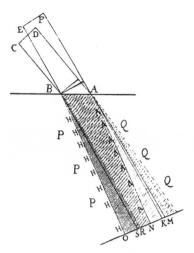

Figure 2.3. From Hooke 1665, Schem. VI., Fig. 4. (Reprinted with the permission of Science Heritage Ltd.)

promoted, or made stronger, having its passage already prepared" for it by the earlier part of the pulse; thus, the red edge of the pulse will spread into the adjacent medium. So in Figure 2.3, without diffusion, the ray EBAF would refract along BONA. But since the red edge diffuses into the adjacent medium, EBAF actually refracts along BOMA (1665, 63). The second factor involved in dispersion is the fact that the physical rays coming from a light source (like the sun) are not parallel. So both physical rays EBAF and CBAD (which might be coming from different edges of the sun) are refracted. Ray CBAD is refracted along the ray with edges BO and Q.

Comments. The light rays of Newton and Hooke have at least four properties in common. Hooke's light rays are *rectilinear* in the absence of mirrors, lenses, prisms and other obstacles. Even the rays that suffer diffusion after having passed through a prism (for example, BOMA in Figure 2.3) are rectilinear. Hooke's light rays travel at *finite velocities;* for example, during refraction light travels faster in one medium than the other. According to Hooke, a physical light ray is a line of light having width along which motion in the ether is propagated (ACDF in Figure 2.2). Since Hooke's ether consists of *irreducibly* tiny particles, some light rays are the irreducibly smallest section of propagating light. Finally, Hooke's light rays are *separable* from other rays. An obstacle can (in principle) separate a ray of light from other collateral and successive rays of light.[3]

Just because Hooke's light rays satisfy the rectilinearity, finite velocity, irreducibility and separability conditions doesn't mean that these assumptions played any role in the *meaning* of the term 'light ray' in Hooke's idiolect. In order to defend this claim, we need to analyze in more detail the way Hooke used this expression in his arguments. And this we will do in a subsequent section (Hooke's Response to Newton).

Newton's First Publication

Newton was convinced to submit a paper to the Royal Society by its secretary, Henry Oldenburg. On 6 February, "The New Theory About Light and Colours" was presented to the Royal Society (1672a). Newton began this paper by setting up the experiment illustrated in Figure 2.4. A circular beam of light is shone through a prism, and the spectrum is then projected to the far wall. Given that the beam entering the prism in Figure 2.4 is *circular,* Newton wondered why, after it was bent by refraction, it does not retain its circular shape. Assuming that the angle of incidence of each ray of white light is the same, then according to the sine law, each ray should be bent the same amount. So it ought to exit the prism at a different angle and retain its circular shape.

Newton offered his explanation for the dispersion: White light

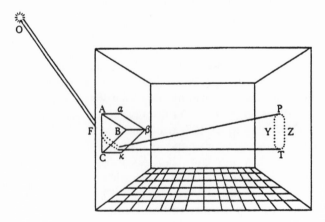

Figure 2.4. From Shapiro 1984, 50. (Reprinted with the permission of Cambridge University Press.)

consists of rays that are bent different amounts by the prism. In order to convince us that white light consists of light rays of different degrees of "refrangibility," Newton described his famous *crucial experiment* (Figure 2.5). "I took two boards and placed one of them close behind the prism at the window, so that the light might pass through a small hole . . ." Newton then took another board with a small hole in it, placed it about 12 feet from the first board, and placed a prism behind the hole in this second board.

> This done, I took the first prism in my hand and turned it to and fro slowly about its axis, so much as to make the several parts of the image cast on the second board successively pass through the hole in it, that I might observe to what places on the wall the second prism would refract them. And I saw by the variation of those places that the light, tending to that end of the image toward which the refraction of the first prism was made, did in the second prism suffer a refraction considerably greater than the light tending to the other end. (1672a, 71)[4]

Here is what happened in Newton's crucial experiment: When the red portion of the spectrum shone through the hole in the second board, it passed through the second prism and was refracted to a particular location on the wall. But when a different colored portion of the spectrum shone through the second board and passed through the second prism, it hit a different portion of the wall. The red light was refracted (bent) less through the second prism than was the yellow light, which was refracted less than the green light, which

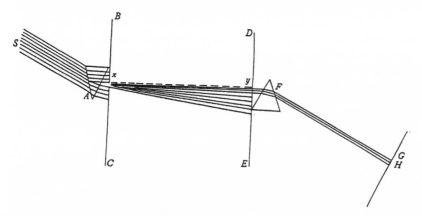

Figure 2.5. From Turnbull 1959, 166. (Reprinted with the permission of Cambridge University Press.)

was refracted less than the blue light, which was refracted less than
the violet light. Newton concluded his discussion of this experiment
by explaining prismatic dispersion.

> And so the true cause of the length of that image was detected to be no
> other than that *light* consists of *rays differently refrangible,* which,
> without any respect to a difference in their incidence were, according to
> their degrees of refrangibility, transmitted toward divers parts of the
> wall. (1672a, 71)

Comments.   The cluster model does a very nice job making sense
of Newton's discussion of his crucial experiment. Newton twice used
boards with holes in them to separate certain rays of light from
others. So it would appear that he was making the separability
assumption. He also made the rectilinearity assumption. A number
of light rays exit the first prism, travel in a straight line path
through the opening in the second board, and into the second prism.
Without this assumption, Newton could not be sure that the rays
entering the second prism had the same angles of incidence. (And
obviously this assumption is important for his conclusion; if the rays
had different angles of incidence, this might explain the different
angles at which the rays exit the prism.)[5] Given just these two
assumptions about rays of light, Newton drew the following conclusion from his observations: different rays of light are bent to different
degrees by the second prism. (Since Newton also claimed that rays of
light are "transmitted," he might also have made the finite velocity
assumption, although it is unnecessary for his argument.)

If we embrace the cluster model, we can interpret Newton's crucial
experiment as employing a concept of light ray that presupposed the
separability and rectilinearity assumptions (and perhaps the finite
velocity assumption). Since the conclusion of the crucial experiment
is that different light rays have different degrees of "refrangibility,"
it is clear that Newton's concept of light ray does not presuppose
what is at issue in this context. The cluster model allows us to
interpret the argument implicit in Newton's crucial experiment so
that it is not circular. In fact, it allows us to interpret Newton's
argument so that it did not assume anything that Hooke would have
found objectionable. For as we have seen, Hooke readily accepted the
separability, rectilinearity and finite velocity assumptions.

The holistic model requires that we interpret Newton's use of the
term 'ray' in his discussion of the crucial experiment in a way that
presupposes the refrangibility assumption. Since this is precisely

what Newton was arguing for, the holistic model requires that we ascribe to Newton a circular argument. Is there some reason why in this particular context Newton would have employed a concept that *assumed* that different light rays have different degrees of refrangibility? I see no reason why we should interpret Newton in this way. In fact, it seems ungenerous to do so. If we have a choice between ascribing to someone a blatantly circular argument or a potentially cogent argument, the charitable thing to do is ascribe to them the potentially cogent argument. But the cluster model has more going for it than a principle of interpretive charity. As we shall see in Newton's rebuttal, Newton repeatedly insisted that he is not employing a concept of light ray that makes any controversial assumptions in this context.

Hooke's Response to Newton

A mere nine days after Newton's paper was presented, Robert Hooke gave his reply to the Royal Society (1672a). Hooke's general strategy was to agree with Newton's experimental findings (while claiming he had already performed them) but argue that they are compatible with and support many other hypotheses about the physical nature of light, including his own. Hooke believed that his account of chromatic dispersion and his wave theory of light could account for Newton's experiments just as well as Newton's hypotheses (see the section on Hooke's *Micrographia* for Hooke's account of chromatic dispersion). So, for example, Hooke agreed with Newton's observations, since he had already made "many hundreds" of those trials and "found them so." But he could not "yet see any undeniable argument to convince me of the certainty" of Newton's hypothesis (1672a, 110). Hooke insisted that his own theory of light could account for Newton's experiments "without any manner of difficulty or straining" (1672a, 111). Indeed, Hooke went on to "assure Mr. Newton" that he could explain all these phenomena "by two or three other" very different hypotheses (1672a, 113).

Hooke argued that since both theories explain all the phenomena, we should accept Hooke's because it is more parsimonious—it does not postulate the existence of as many different types of entities. On Newton's theory, each color is basic "amongst which are yellow, green, violet, purple, orange, and so on, and an indefinite number of intermediate gradations; I cannot assent thereunto, as supposing it wholly useless to multiply entities without necessity" (Hooke 1672a, 113).

Comments. In order for Hooke to agree (or at least believe he was agreeing) with Newton's experimental findings, Hooke must have believed that there is a concept of light ray that they could both use without inconsistency. If the holistic model is accurate, then Hooke was wrong. But what is perhaps worse, Hooke was completely blind to Newton's use of a concept that begged the question against his theory. It seems hard to believe that Hooke could have been oblivious to an argument whose circularity is so blatant.

The cluster model does not have these consequences. Consider a passage in which Hooke agreed with the results of Newton's crucial experiment but not his explanation of it.

> ... Mr. Newton alleges that as the rays of light differ in refrangibility, so they differ in their disposition to exhibit this or that color: with which I do in the main agree ... [T]he ray by refraction is, as it were, split ... [and] the one side, namely that which is most refracted gives blue, and that which is least gives red. (Hooke 1672a, 112)

What does Hooke mean by 'ray' in this passage? The entire visible ray of light that passes through a prism. After the refraction, one side of the ray "gives blue" and the other side "gives red." Hooke was clearly not making the irreducibility assumption here since the 'ray' in question can be separated into smaller portions. All he seemed to be assuming is that light rays are visible beams of light that travel in straight lines (in absence of obstacles). And these are assumptions he could share with Newton.

Newton's Rebuttal

Initially, Newton was unruffled by the patronizing tone of Hooke's reply. But within four months, Newton's attitude had changed—his published reply to Hooke was harsh and derisive (Newton 1672b; see Westfall 1963 for a fascinating chronology of this change). Newton's angry rebuttal was three pronged. First, Newton offered further experiments that allegedly support his theory of colors but not Hooke's. Second, Newton claimed not to have embraced with any confidence the particle theory of light and he insisted that his arguments can be made without assuming anything about the physical substructure of light.

> ... I knew that the properties which I declared of light were in some measure capable of being explicated not only by that, but by many other

mechanical hypotheses. And therefore I chose to decline them all, & speak of light in general terms, considering it abstractedly as something or other propagated every way in straight lines from luminous bodies, without determining what that thing is, whether a confused mixture of difform qualities, or modes of bodies, or of bodies themselves, or of any virtues powers or beings whatsoever. (1672b, 174)

The third prong of Newton's rebuttal was to argue that Hooke's wave theory of light is "impossible" and "not only *insufficient,* but in some respects *unintelligible*" (emphasis in original, 1672b, 175–6). This is a charge Newton would make throughout his career. Newton argued that it is "impossible . . . that the waves or vibrations of any fluid can like the rays of light be propagated in straight lines, without a continual & very extravagant spreading & bending every way into the quiescent medium where they are terminated by it" (1672b, 175). Newton's argument is that if light were a wave phenomenon, when light waves pass a barrier, they should, like water and sound waves, propagate into the still medium (shadow) behind the barrier. But, according to Newton, they do not. So light cannot be a wave phenomenon.

I will ignore Hooke's defense of his position in his letter to Lord Brouncker, which Newton probably never saw (Hooke 1672b). It is safe to say, however, that by the end of 1672, neither man had altered his views about chromatic dispersion or the nature of light.

Comments. Throughout his career, Newton insisted that he did not employ a concept of light ray that made controversial assumptions (especially the particulate or refrangibility assumptions). But if we accept the holistic model, it follows that Newton was self-deceived about the nature of his concept. How did this happen? It seems incumbent upon the defender of the holistic model to explain how Newton and Hooke could have been so deluded about what they were talking about.

The cluster model provides a more natural interpretation of Newton's response. It does not require us to believe that Newton was systematically misled about the nature of his concepts. When Newton insists that he "speak[s] of light in general terms, considering it abstractedly as something or other propagated every way in straight lines from luminous bodies, without determining what that thing is," the defender of the cluster model can take him at his word.

In general, the cluster model offers smoother, more plausible interpretations of Newton and Hooke. The holistic model makes

Newton and Hooke appear unaware about what they were talking about. It also renders their arguments circular. Of course, sometimes people give dumb arguments. But it does seem incumbent upon defenders of the holistic model to explain why Newton and Hooke, who by anyone's standards were no dummies, could have proposed such poor arguments. The holistic model also seems ungenerous in its interpretations: Why attribute to anybody a circular argument when there is a perfectly adequate noncircular argument available that the person seems to be insisting upon?

Two more general points in favor of the cluster model. First, the cluster model respects our intuition that in different contexts, we can use the same phrase to mean quite different things. We can convince ourselves of this by thinking about our own case. If we want to describe how light rays exit a prism, we can use the term 'light ray' to mean no more than visible lines of light. But if we want to give a Newtonian explanation of refraction, we can use the term 'light ray' to mean a particle travelling at a certain speed. If *we* can express different concepts in different situations, it seems reasonable to suppose Newton could have done so too. The second point in favor of the cluster model is that it is more powerful as an interpretive device than the holistic model: any interpretation of a passage available to the holistic model is also available to the cluster model, but not vice versa. Both models can account for Newton (or Hooke) employing a controversial concept. But only the cluster model can account for them employing a noncontroversial concept.

ARGUMENTS FOR THE HOLISTIC MODEL

I have argued that the holistic model does a poor job making sense of the Newton-Hooke controversy. So why would anyone accept it? The argument for the holistic model begins with the definition of 'light ray' Newton gives in the *Opticks*.

By the rays of light I understand its least parts, and those as well successive in the same lines, as contemporary in several lines. For it is manifest that light consists of parts, both successive and contemporary; because in the same place you may stop that which comes one moment, and let pass that which comes presently after; and in the same time you may stop it in any one place, and let it pass in any other. For that part of light which is stopped cannot be the same with that which is let pass. The least light or part of light, which may be stopped alone without the rest of the light, or propagated alone, or do or suffer any thing alone, which the rest of the light doth not or suffers not, I call a ray of light. (1730, 2)

Buchwald claims this definition "amounts to defining a ray as an atom of light" (1989, 7). And Shapiro maintains that "this was not unintentional on Newton's part" (1975, 209).

The argument for this interpretation is spelled out in most detail by A. I. Sabra (1967). Newton's definition says that there is (in principle) some "least" part of light that can be separated from other parts of light. Sabra argues that this assumption implies that light rays are particles.

> But what does he mean by the 'least' or 'indefinitely small' parts of light? For what is being made indefinitely smaller and smaller is, in the first operation, a region of space and, in the second, an interval of time. Newton is obviously making the assumption that this double process by which the beam is being chopped both spacially and temporally may be imagined to come to an end before the hole is completely closed. Thus by making the hole narrow enough only those rays coming successively in the same line will be let through. And, further, by making the interval during which the hole is open sufficiently small, only one of these rays will escape. It was this assumption which made Newton's critics suspect that his rays . . . were simply the corpuscles, in spite of Newton's refusal to attach the proper label upon them. (289)

There are two reasons to reject this argument. First, it grants too much importance to Newton's self-conscious definition. Even if Newton's definition presupposes that light is particulate, it does not follow that Newton used the term 'light ray' in this way throughout his optical writings. Second, when properly understood, this definition simply does not presuppose a particle theory of light. Newton's definition gives us a method for isolating a ray of light: that "least part of light" which passes through the temporarily unobstructed hole is a ray of light. But because Newton believed that that thing is a moving particle, Sabra concludes that Newton's definition presupposes that light is particulate. But this does not necessarily follow. Although Newton believed that his definition *refers* to a moving particle of light, it does not follow that this is part of the *meaning* of 'light ray'. Indeed, the cluster model offers a competing interpretation: Newton's definition need only presuppose that there is some *irreducibly smallest section of propagating light*. This might be a moving particle or some kind of motion transferred along a line of ether particles.

Can we choose between these interpretations of Newton's definition? Yes. If Newton were assuming that light is particulate, his definition should commit him to a view about how light travels. In particular, it should commit him to the view that the irreducibly smallest sections of light actually travel through space. But in an unpublished reprint of his first paper, Newton explicitly declared that he makes no assumptions about whether light rays are "trajected" or whether they "propagate motion from one to another in right lines." Newton stated that whether light rays are particulate or wave-like,

> ... light is equally a body or the action of a body in both cases. If you call its' rays the bodies trajected in the former [particle theory] case, then in the latter [wave theory] case they are the bodies which propagate motion from one to another in right lines till the last strike the sense. The only difference is, that in one case a ray is but one body, in the other many. (Cited in Cohen 1958, 365)

Since Newton explicitly insisted that his concept of light ray is silent about how light moves, Newton's definition in the *Opticks* only committed him to the view that there is some irreducibly smallest section of propagating light. And this is perfectly consistent with Hooke's wave theory of light. In fact, it is perfectly consistent with all the major wave theories of light Newton knew about, including those of Descartes, Hobbes, and Huygens. All of these wave theories postulated ethers that were made out of particles (for a lucid summary of these theories, see Shapiro 1974).

CONCLUSION

If the analysis given here of the Newton-Hooke controversy is roughly on track, then we must reject an entire class of philosophical accounts of scientific change. A number of philosophers, including Kuhn, Feyerabend and Churchland, have argued that deeply different and competing scientific theories inevitably employ radically different, and even incommensurable, concepts. A number of historians, including Sabra, Shapiro, and Buchwald, have analyzed the 1672 Newton-Hooke controversy in this way. But in reality, the Newton-Hooke case study shows something quite different: Even when two scientists embrace dissimilar theories and engage in a bitter controversy, it is possible for them to use the same concepts to describe the parts of the world in dispute. Communication break-downs across deep theoretical divides are not inevitable.[6]

NOTES

1. Ronchi and Sabra argue that throughout Newton's career the particulate assumption was a necessary condition for something being a light ray; Shapiro argues that the homogeneity, heterogeneity and refrangibility assumptions were necessary conditions, but that the particulate assumption was necessary only later in Newton's career. Buchwald agrees with Shapiro about the particulate assumption.

2. While this seems to be a particularly outlandish consequence of holism, we should note that proponents of semantic holism tend to deny that there are any analytic statements (see for example, Churchland 1979). Without getting into this quagmire, can we make sense of the holist's claim? I think so. Consider: many dictionaries define whales as mammals. And it would not be surprising if many people's concept of whale involved *being a mammal*. For these people, the statement 'All whales are mammals' would be analytic in the same sense the holist is claiming that 'Light rays consist of particles' is analytic for Newton. 'Whales are mammals' is not analytic because someone stipulated that this was how we would use the term 'whale'. We empirically discovered that whales are mammals and (for various reasons) this has become part of our concept of whale. It is logically possible that we might discover, to our great surprise, that whales are not really mammals. I think the holist should best be understood as claiming that theoretical statements are analytic in the same sense that 'Whales are mammals' is analytic. And although we may decide that this view is confused, it is not obviously absurd.

3. One might argue as follows that the rectilinearity and separability assumption together are incompatible with wave theories of light: "These two assumptions imply that after a single ray passes by an obstacle, it will continue in a straight line. But wave theories of light do not have this consequence. When a single particle on a wave front passes by an obstacle, it will produce a (secondary) wave front (in accordance with Huygens' principle)."

In order to respond to this objection, we must recognize that seventeenth-century wave theorists employed a concept of *light ray* and of *light wave*. While Hooke and Huygens believed that light waves (or pulses) spread out when passing by certain obstacles, they also believed that light rays are rectilinear, even when a single ray passes by an obstacle (see Shapiro 1974). The point to keep in mind is that wave and particle theorists were capable of sharing and employing concepts of *light ray*.

4. Throughout this paper, I have made minor spelling and punctuation changes in the original letters written by Newton and Hooke that are reproduced in Turnbull (1959) to make the material more natural for modern readers.

5. As a matter of fact, the rays did enter the second prism with slightly different angles of incidence. For a very interesting discussion of this point, see Ronald Laymon (1978).

6. I would like to thank Laura Snyder and especially Peter Achinstein for very helpful comments on earlier versions of this paper.

REFERENCES

Achinstein, Peter. 1968. *Concepts of Science*. Baltimore, Maryland: The Johns Hopkins University Press.

Buchwald, Jed Z. 1989. *The Rise of the Wave Theory of Light*. Chicago: University of Chicago Press.

Churchland, Paul M. 1979. *Scientific Realism and the Plasticity of Mind*. Cambridge: Cambridge University Press.

Cohen, I. B. 1958. "Versions of Isaac Newton's first published paper," *Archives internationales d'histoire des sciences* 11: 357–375.

Feyerabend, Paul. 1975. *Against Method*. London: New Left Books.

Hooke, Robert. 1665. *Micrographia*. Reprinted in 1987 in Lincolnwood, Illinois: Science Heritage Limited.

———. 1672a. "Robert Hooke to Henry Oldenburg, Feb 15, 1671/2" in Turnbull, 110–114.

———. 1672b. "Robert Hooke to Lord Brouncker, June, 1672" in Turnbull, 198–205.

Johnson, B. K. 1960. *Optics and Optical Instruments*. New York: Dover Publications.

Kuhn, T. S. 1970. *The Structure of Scientific Revolutions*. Chicago: The University of Chicago Press.

Laymon, Ronald. 1978. "Newton's *Experimentum Crucis* and the Logic of Idealization and Theory Refutation," *Studies in the History and Philosophy of Science* 9: 51–77.

Newton, Isaac. 1672a. "The New Theory about Light and Colors," in Thayer, 68–81.

————. 1672b. "Isaac Newton to Henry Oldenburg, June 11, 1672" in Turnbull, 171–193.

————. 1730. *Opticks,* 4th edition. Reprinted in 1979 in New York: Dover Publications.

Ronchi, Vasco. 1970. *The Nature of Light: An Historical Survey.* Cambridge: Harvard University Press.

Sabra, A. I. 1967. *Theories of Light from Descartes to Newton.* London: Oldbourne Press.

Searle, John. 1958. "Proper Names," *Mind* 67: 166–173.

Shapiro, Alan. 1974. "Kinematic Optics: A Study of the Wave Theory of Light in the Seventeenth Century," *Archives for the History of the Exact Sciences* 11: 134–266.

————. 1975. "Newton's Definition of a Light Ray and the Diffusion Theories of Chromatic Dispersion," *Isis* 66: 194–210.

————, ed. 1984. *The Optical Papers of Isaac Newton, Vol. I.* Cambridge: Cambridge University Press.

Thayer, H. S., ed. 1953. *Newton's Philosophy of Nature: Selections from his Writings.* New York: Hafner Press.

Turnbull, H. W., ed. 1959. *The Correspondence of Isaac Newton.* Cambridge: Cambridge University Press.

Westfall, Richard. 1963. "Newton's Reply to Hooke and the Theory of Colors," *Isis* 54: 82–96.

Wittgenstein, Ludwig. 1953. *Philosophical Investigations,* trans. G. E. M. Anscombe. Oxford: Basil Blackwell.

3

EXPERIMENTAL SKILLS AND EXPERIMENT APPRAISAL

Xiang Chen

ABSTRACT

Traditional philosophy of science believes that scientists can achieve agreement on every experimental result provided it can be replicated in an appropriate way, that is, reproducible with the same experimental arrangement and procedure. By analyzing the role of skills in experiment appraisal, I explain why in fact scientists do not always have consensus on experimental results despite their replication attempts. Based on a detailed analysis of a historical case, I argue that experiment replications inevitably involve a process of skill-transference, which is frequently not articulated in linguistic discourses. Hence, it is very difficult to make identical replications if experimental reports are the only resources. Furthermore, I argue that, because transferred skills have to be integrated with scientists' prior experience, skill-transference is sensitive to contextual factors, which can prevent scientists from reaching consensus on experimental results by influencing the effectiveness of communication in experiment appraisal.

INTRODUCTION

Every student of science agrees that experiment is the foundation of theory testing, because experiment is supposed to supply objective knowledge of the world. However, experimental results themselves are not unproblematic. Most experimental instruments, procedures, and findings now widely accepted as reliable have experienced a period in which their legitimacy was controversial. Even after these instruments, procedures, and findings become conventional, their

45

legitimacy may later be challenged under new circumstances, especially in scientific debates. Hence, not every experimental result can be qualified as objective knowledge. Whether an experiment, especially a newly designed one, can provide objective knowledge about the world is a question that requires careful examination. Experiment appraisal, that is, evaluating the legitimacy, the reliability, or the accuracy of experimental instruments, procedures, and findings, is an important topic for the philosophy of science.

Some contemporary philosophers of science have addressed the issue of experiment appraisal. Karl Popper, for example, notes that an experimental result must satisfy a couple requirements in order to be qualified as objective knowledge. First, this result must involve a genuine physical effect, which is observable. To be more specific, it should be an observable effect "occurring in a certain individual region of space and time," or involving "position and movement of macroscopic physical bodies"[1] (1959, 103).

Second, and more important, an acceptable experimental result should be reproducible. Popper maintains that "[w]e do not take even our observations quite seriously, or accept them as scientific observations, until we have repeated and tested them. Only by such repetitions can we convince ourselves that we are not dealing with a mere isolated 'coincidence', but with events which, on account of their regularity and reproducibility, are in principle inter-subjectively testable" (Ibid., 45).

However, Popper also realizes that a reproducible result may not need to be actually reproduced (Ibid., 87). The key to demonstrating the reproducibility of an experimental result, according to Popper, is to provide clearly written instructions for its replication, so that the result "can be regularly reproduced by anyone who carries out the appropriate experiment in the way prescribed" (Ibid., 45). He thus recommends that experimental processes should be expressed in clearly written descriptions. Those who conduct an original experiment should present the experiment by describing the experimental arrangement in detail so that anyone with relevant techniques can replicate it. Those who have doubts about the original experiment should construct a counter-experiment with contradictory results, and publish instructions telling others how to repeat their new experiment (Ibid., 99).

If scientists follow this methodological guideline carefully, and if what they are dealing with is a physical effect involving position and movement of macroscopic bodies, they should be able to reach agreement about the experimental result, or at least to determine whether there is any difference in their experimental arrangements or operations. Otherwise, Popper says, language would no longer be "a means of universal communication" (*Ibid.*, 104).

However, the results of some recent studies of scientific experiments suggest that replication attempts, even those that strictly follow Popper's methodological guidelines, do not always produce agreement among scientists about experimental results. Based on detailed analyses of experimental discoveries in contemporary physics, for example, Franklin and Howson report that experiment replications are neither necessary nor sufficient for the validation of experimental results (Franklin and Howson 1988, 426). Also, based on interviews with a group of biochemists, Mulkay and Gilbert note that scientists frequently have different conceptions of what a proper experiment replication should be, and that replication attempts may not bring about agreement among scientists though the experimental results they are dealing with are observable physical effects (Mulkay and Gilbert 1986, 22).

The purpose of this paper is to examine the complexities involved in experiment appraisal, and to explore some of the fundamental features of experiment replications. In the following sections, I first illustrate the complexities in experiment appraisal by analyzing a historical case: the debate over the result of a prismatic interference experiment in the early 1830s. The main issue of this debate concerned what exactly happened in the prismatic interference experiment. Even though the experiment produced an observable physical effect, conflicting reports about the experimental result per se still existed after a series of replication attempts. This historical episode vividly shows that replicating an experiment "in the way prescribed" cannot always verify the experimental result even if it is a genuine physical effect.

I then explore one of the crucial features of experiment appraisal that has been underestimated by Popper: the involvement of experimental skills. I argue that experiment replications inevitably require a process of skill-transference, which is frequently not articulate in linguistic descriptions. We should not expect that experimental processes can be described in clearly written instructions so that others can reproduce experiments "in the way prescribed." Moreover, I argue that those transferred skills have to be integrated with people's prior practices or experiences. Thus, even if clearly written instructions have been given, even if these instructions have been carefully followed, and even if the experimental results are genuine physical effects, scientists still may not be able to reproduce the same experimental result because of their different prior practices or experiences.

THE DEBATE ON THE PRISMATIC
INTERFERENCE EXPERIMENT

In Britain, the early 1830s was a critical period for the development of optics. Since Newton's endorsement in the late seventeenth century, the particle theory of light, which claimed that light is composed of tiny particles, had dominated the field of optics in Britain for more than a hundred years. During this period, the wave theory of light, which regarded light as waves, was very unpopular. The dominance of the particle theory, however, became shaky at the beginning of the nineteenth century. In the late 1820s, a group of British scientists, most of them Cambridge-trained physicists, adopted the wave theory. Beginning in 1830, these newly committed wave theorists started to publish their results, both theoretical and experimental. A heated particle-wave debate then began.

In 1832, Baden Powell, Savilian Professor of geometry at Oxford and a committed wave theorist, published an article in *Philosophical Magazine* on several experiments about diffraction and interference (Powell 1832a). One of the experiments that Powell described in detail was originally proposed by Augustin Fresnel. This was an experiment using two plane glasses inclined at a very large angle to demonstrate the phenomenon of interference by reflection. Powell repeated this experiment with some modifications. In addition to having two plane glasses inclined at a very large angle as Fresnel did, Powell placed a prism in front of the glasses, in the position where the two reflected rays were supposed to intersect (Figure 3.1). Using sunlight as the light source, he found that, after being refracted by the prism, the two reflected rays continued to produce interference fringes—a series of parallel alternating light-and-dark lines. He also found that the pattern and the positions of the interference fringes did not change after the interception by the prism. Powell believed that the results of this prismatic interference were entirely consistent with the wave theory.

Powell's experiment on prismatic interference drew the attention of Richard Potter, an amateur physicist at the time.[2] Although he was a merchant at Manchester, Potter devoted his leisure time to the study of optics, conducting experiments to measure the reflective power of mirrors. Since he found that neither the particle theory nor the wave theory was able to explain the experimental results he

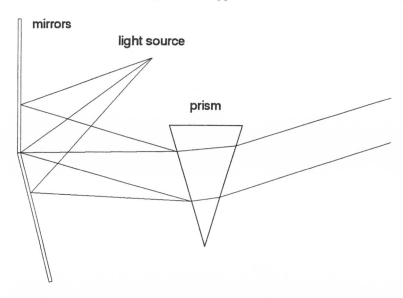

Figure 3.1 Powell's prismatic interference experiment

obtained, Potter did not commit himself to either theoretical tradition in his early optical researches.

   After reading Powell's article, Potter replicated the prismatic
interference experiment. Instead of using sunlight as the source,
Potter employed homogeneous light produced by a colored solution.
He observed that some portions of the reflected rays, which should
have interfered without the prism, did not interfere after being
refracted by the prism. On the other hand, he found that interference
took place between other portions of the reflected rays. Using an
eyeglass to observe the interference fringes directly, Potter found
that the interference fringes moved toward the thick side of the
prism when he withdrew his eye and the eyeglass farther from the
prism. In February 1833, Potter published a paper in *Philosophical
Magazine,* reporting his experimental findings. As shown in Figure
3.2, Powell had reported that the interference fringes produced after
the refraction by the prism were unchanged, and the central band of
the interference fringes was still on the line *mn.* However, Potter
reported that different portions of the reflected rays were involved in
the interference, and that the central band of the interference fringes
was on a new line *pq* (Potter 1833a, 82).

   Potter also found that this experiment on prismatic interference

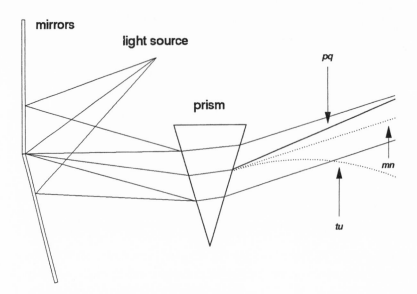

Figure 3.2 The result of Potter's replication

could be used to determine the velocity of light in refractive media. The positions of the interference fringes in this experiment were determined by the path differences of the intersecting rays. These path differences were affected by the prism because rays of light changed their velocities in refractive media. Hence, the velocity of light in the prism could be calculated based upon the position of the central band of the interference fringes. Since the two rival theories of light had different predictions of the velocity of light in refractive media, a test of these theories could be made by comparing their predictions with the measures.

   The particle theory assumed that light moved with an increased velocity when passing through refractive media, in a direct ratio to their refractive indices. According to this assumption, Potter found, the central band of the interference fringes in his experiment ought to be seen along the line *tu* in Figure 3.2, which was far from the facts shown by the experiment. On the other hand, the wave theory assumed that light traveled with a decreased velocity in refractive media, in an inverse ratio to their refractive indices. According to this assumption, Potter demonstrated, the central band of the interference fringes in this experiment should coincide with the intermediate line *mn,* which was still not compatible with the experimental results, although better than the particle theory's prediction. Therefore, neither the particle theory nor the wave theory of light gave a correct prediction of the velocity of light in refractive media. These experimental results, Potter claimed, constituted a fatal objection to both theories of light (*Ibid.*, 94).

Potter's attack prompted strong reactions from the wave camp, including one from George Airy. As Lucasian Professor at Cambridge, Airy was one of the most influential figures among the wave theorists in the early 1830s, and had been known to fiercely counterattack every challenge from the particle camp. Airy published a comment on Potter's experiment in *Philosophical Magazine,* just one month after the appearance of Potter's article. In his remarks, Airy first cast doubt on one of the most important experimental conditions in Potter's work—the light source. Airy insisted that Potter must not have used homogeneous light as the source in his experiment. Airy listed two reasons to support his allegation. First, interference by reflection required a light source with very high intensity, but so far all homogeneous sources could only produce very faint light. Second, if homogeneous light had been used in Potter's experiment, Airy reasoned, it would have produced a series of bright and dark bars with equal intensity, and no one could have determined where the center of the fringes was (Airy 1833a, 164, 162).

If the light source was not homogeneous but heterogeneous, Airy argued, then the center of the fringes was not at the point where the two intersecting rays had equal paths—the line *mn* in Figure 3.2. Airy emphasized that his analysis of the positions of the interference fringes was theoretically neutral, having nothing to do with assumptions about the nature of light. According to Airy, if a heterogenous light source was used, each homogeneous ray composing the reflected heterogeneous light would produce its own group of bars. Due to the impact of the prism, the bars produced by each color would have different breadths and different displacements moving slightly toward the thick end of the prism. When these different groups of bars coincided with each other, they constituted the center of the fringes with a displacement toward the thick end of the prism, although the group of interference fringes as a whole actually did not move (*Ibid.*, 162–4).

Airy realized that the phenomenon he said he could explain was not identical with the one Potter claimed he had observed in the experiment. Potter said that he had seen the displacement of a *group* of the interference fringes as a whole, while Airy only accounted for the shift of the *center* of the fringes. But Airy insisted that Potter's observation must be wrong because of an inappropriate observation technique he employed. To illustrate this point, Airy presented to his readers an "instructive experiment." This was also an experiment on interference by reflection, in which two pieces of glass were connected with hinges. By slightly inclining one piece of glass while fixing the other, the center of the interference fringes would move while the position of the group of fringes as a whole remained unchanged. However, Airy noted, if the experimenter had not been continuously observing the change of the fringes, for example, if he left the eyeglass to adjust the angle between the glasses, he might not be able to distinguish the differences between a shift of the center and a move of the group as a whole, because he could not tell whether the rest of the fringes really had moved. A continuous observation, thus, was a key for achieving reliable results. Airy suspected that, without any discussion of this issue in his experimental report, Potter must have been unaware of the problem and not been continuously observing the change of the fringes when he withdrew his eye and the eyeglass from the prism. Therefore, Potter's observation was not reliable (*Ibid.,* 164–5).

Potter was very unhappy with Airy's remarks. He immediately published a reply in the 1833 *Philosophical Magazine,* in which he complained that Airy's analysis of his experiment had completely missed the point. In response to Airy's charge about his experimental setting, Potter provided details about the light source he had used in the experiment. It was the red light produced by a solution of "iodine in hydriodic acid," which gave much purer and more intense light than red glasses did. Even according to the standard adopted by wave theorists like Fresnel, Potter claimed, the light source in his experiments was satisfactorily homogeneous (Potter 1833b, 276–7).

Potter also held that the observation techniques he used were reliable and would not create the confusion that Airy had described. One of the advantages of his techniques, Potter claimed, consisted in the use of a reference to show the displacement of the interference fringes. This was a group of diffracted lines caused by the edge of one of the glass mirrors during the observation process. To illustrate this point, Potter presented a diagram (Figure 3.3), in which lines *ef* represented the diffracted fringes produced by the edge of the mirror, and *ab* and *cd* were the different positions of the whole interference fringes he observed at different distances from the prism. By introducing this reference, Potter said that he could be certain about the movement of the interference fringes as a whole, and did not commit the observational mistake that Airy had suggested (*Ibid.,* 287).

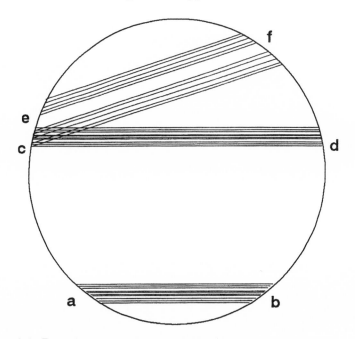

Figure 3.3 Potter's observations of the fringes

The most powerful defense Potter presented, however, was his announcement that he had successfully replicated his experiment in front of Powell. He claimed that he had repeatedly replicated his experiment at Powell's residence in June 1832 (*Ibid.*). There is not further evidence to verify Potter's replication attempts. But from a paper Powell published in December 1832, it is evident that Powell had known of Potter's experimental results, and, surprisingly, adopted a very positive attitude toward Potter's work (Powell 1832b, 436).

The dispute between Potter and Airy finally centered on a very simple question: What had actually happened in these experiments? Or, more specifically, had the group of interference fringes as a whole really moved in these experiments or had they not? Potter's claim concerning his successful replications in front of Powell forced Airy to replicate the experiment of prismatic interference. In his replication, Airy used a new observation method to determine the displacements of the fringes. His new idea consisted in using an eyeglass with a wire fixed in its focus, both attached to a slide on a bar. By proper adjustment of the bar's direction, Airy was able to keep the image of the wire focused upon one of the fringes when he looked through the eyeglass, even though the distance between the eyeglass and the prism varied. With this method, Airy reported that, while the fringe under the wire shifted only a half of its breadth, the center of the fringes had gradually moved through a distance of twelve double fringes (Airy 1833b, 451).

Airy's new observation device did not convince Potter. After reading Airy's report of the replication, Potter immediately complained that Airy did not give sufficient information to enable others to verify the result. He pointed out that Airy's description of his new observation device was not sufficient for further replications, unless the angle between the bar and the incident rays, together with other data, were known. Potter also charged that Airy's observation device created unnecessary "intricacy," because it introduced a new object, a wire fixed in the focus of the eyeglass, as the reference. This "intricacy," according to Potter, could be avoided by using a reference that had been given by the experimental arrangement. A reliable observation can be obtained by measuring the position change of the interference fringes with respect to the diffracted fringes caused by the edge of one of the mirrors (Potter 1833c, 333). For these reasons Potter concluded that Airy's replication could not be reliable. Thus, after several rounds of exchanges, Potter and Airy still did not reach agreement on what really happened in these experiments. Specifically, they simply did not agree with each other on whether the position of the group of fringes as a whole moved when they were observed from different distances.

The attempts to replicate the prismatic interference experiment, which produced more than ten experimental reports from Powell, Potter, Airy and others between 1832 and 1833, did not yield any agreement. On the one hand, the wave theorists in the debate were confident that, through their replications, the problem had been successfully solved by the wave theory. Airy even predicted that, if Potter continued to study this subject, he would very soon become a wave theorist (Airy 1833a, 167). On the other hand, Potter regarded the results in his replications as a solid evidence against the wave theory, and claimed that it was harder and harder for him to accept the wave theory (1833b, 277).

These completely opposite judgments stemmed from Airy's and Potter's different observations of what really happened in the experiments. The discrepancy could perhaps have been resolved through performing the experiment in front of the two scientists. But in a letter to William Hamilton on April 1833, Airy expressed his reluctance to continue the debate with Potter or to verify the experimental findings in question.[3] One reason suggested for Airy's retreat was that the debate had become too personal.[4] However, a more plausible reason was that Airy just did not have an interest in meeting with Potter. There were great differences between Airy and Potter in terms of their social and intellectual status. In the early 1830s Airy had been one of the most successful and prestigious scientists in Britain. Potter, on the other hand, was an unknown amateur who had no formal training in science. Such differences could create a barrier to a face-to-face meeting between Potter and Airy, which might have helped them determine the details of their experimental settings and resolve their differences.

EXPERIMENTAL SKILLS AND CONTEXTUAL FACTORS

The debate between Potter and Airy on the prismatic interference experiment indicates that the process of experiment appraisal is much more complicated than what Karl Popper has described. Although the prismatic interference experiment did produce a real physical effect (interference fringes involved only position and movement of macroscopic bodies), scientists failed to reach agreement about what this physical effect was despite several replication attempts. Potter and Airy simply did not agree with each other on what really happened in the experiment: the former insisted that he saw the displacement of the group of the interference fringes while the latter maintained that only the center of the fringes shifted.

The unsettled debate on the experiment raises some important questions. Why did Potter and Airy fail to reach agreement on the result of the experiment, which was a real physical effect? Or, if they were in fact dealing with different physical effects by conducting different experiments, why did they fail to detect their differences and resolve the debate? The deadlock between Potter and Airy suggests that they had experienced a communication failure that hindered them from reaching consensus on the experimental result. If so, what were the factors that caused the communication failure?

The peculiar inconclusiveness of experiment appraisal in the debate on the prismatic interference experiment might partly result from the conventional style of reporting and representing experimental findings. In early nineteenth-century Britain, there was no standard format for reporting optical experiments. Most experimental reports on optics published in academic journals were relatively

simple, usually lacking detailed descriptions of experimental procedures, instruments, and results. This was particularly true for those
publishing in *Philosophical Magazine*. Unlike *Philosophical Transactions, Philosophical Magazine* provided only a very limited space
for publications. Within an average length of three to five pages, it
was quite difficult to portray an experiment in detail or to provide
the necessary information for experiment replications.

Moreover, scientists in the early nineteenth century lacked adequate techniques to reproduce optical images in their experimental
reports. Before 1860 when photographic techniques became available for book or journal illustrations, scientists in optics were limited
to sketches and engraved diagrams. But these techniques could not
accurately represent the details of optical images, especially the
variations of the intensity of light. In the debate on the prismatic
interference experiment, less confusion would have been created if
Airy had been able to reproduce his observation in his report with an
accurate illustrative technique, rather than just giving a verbal
description. For example, Airy described his observation as follows:
"[O]n receding from the prism, the *fringes* remain stationary; while
. . . the *centre of fringes* passes gradually and rapidly from the centre
of the mixture of light to its border" (Airy 1833b, 451; original
emphasis). According to this description , the center of the fringes
experienced a spatial displacement, moving from one location to
another. However, William Whewell, who agreed with Airy on the
experimental result, had a different description of the same phenomenon. After witnessing Airy's experiment, Whewell wrote down his
observation as follows: "As you withdraw the eyepiece, you see the
bars, not move, but *grow* on one side and *dim* on the other so that the
centre shifts" (Hankins 1980, 150; original emphasis). According to
Whewell's description, there was no spatial movement but rather
changes of the intensity of light. This confusion could have been
eliminated by using an appropriate technique such as photography
that could capture the optical phenomenon in detail and correctly
present it to the readers.

These limitations increased the difficulties in experiment replications, if one had only the information from published experimental
reports. To complete the replication process for experiment appraisal, intensive communication, especially informal exchanges,
between scientists was necessary. In terms of their functions in
experiment appraisal, there were significant differences between
formal communication (experimental reports and published replies
or comments) and informal communication (private conversations

and private correspondence). In the debate on the prismatic interference experiment, formal communication might be able to assure those who did not have direct experience of the experiment in question and did not intend to replicate it that its result was reliable. However, it was not enough to persuade those who had been directly involved in the debate, because the experimental reports and published replies did not supply the detailed information sufficient for experiment replications. Only informal communication that aimed at information exchange between relevant scientists could complete the process of replication. These informal exchanges, however, depended upon a series of contextual factors. As indicated above, the differences in intellectual and social status between Potter and Airy might have prevented them from further private communication, even though Airy was willing to reply publicly to Potter.

The format of experimental reports, the techniques of presenting optical images, and the intellectual and social status of scientists were the contextual factors that contributed to a communication failure between Potter and Airy in their appraisals of the prismatic interference experiment. Our historical episode clearly indicates that contextual factors played a significant role in the evaluation of the prismatic interference experiment. However, was the involvement of contextual factors in this historical case merely contingent, or inevitable in the sense that it reflected an essential feature of experiment appraisal? If the answer is the latter, then a related question also should be asked: how are contextual factors in general involved in the process of experiment replication?

One way to answer these questions is to examine the distinct characteristics of experiment appraisal. For a long time, philosophers have recognized that there are some fundamental differences between two kinds of intelligent activities: knowing how and knowing that. As noted by Gilbert Ryle more than forty years ago, "there are many classes of performances in which intelligence is displayed, but the rules or criteria of which are unformulated" (1949, 30). Examples of these performances, or the activities of *knowing how,* include a wit who knows how to make good jokes and how to detect bad ones, but cannot tell us or himself any recipes for doing so; or, a well-trained sailor who can tie complex knots and discern if someone else is tying them correctly, but who is probably incapable of describing in words how the knots should be tied. By contrast with *knowing that* (in which intelligent operations involve the observance of rules), *knowing how* involves the intelligent activities in such a way that "[e]fficient practice precedes the theory of it; . . . Some

intelligent performances are not controlled by any anterior acknowl-
edgments of the principles applied to them" (*Ibid.*).

Another writer who comments on the differences between knowing how and knowing that is Michael Polanyi. He labels the products
of knowing how as tacit knowledge or skills. One example Polanyi
uses to illustrate the characteristics of tacit knowledge or skills is
the practice of cycling. In this case, the major task for a cyclist is to
keep balance. According to our knowledge of physics, we know that,
in order to compensate for a given angle of imbalance, we must take
a curve on the side of the imbalance, of which the radius should be
proportional to the square of the velocity divided by the tangent of
the angle of imbalance. We may write down this requirement in the
form of a rule, but learning this rule certainly does not make one
know how to cycle. In fact, the majority of cyclists would not be able
to describe in words this rule, although they know quite well how to
keep their balance (Polanyi 1969, 144). Explicit knowledge of rules
in this case may be completely ineffectual. On the other hand,
Polanyi does not exclude the possibility that some cyclists may
improve their skills in cycling by studying the rules written down in
manuals or taking instructions of experts. But he insists that, when
they come to action, they have to "reintegrate" this explicit knowledge of rules with their prior performances (Polanyi 1966, 11). This
knowledge of rules has to be reapplied in a new situation, one in
which the knowledge of rules itself does not specify its application
conditions. Hence, the key to success in these cases is not the
knowledge of rules but people's prior practice or experience that
shapes the applications of rules.

The process of experiment appraisal involves a variety of activities
that clearly belong to knowing how rather than knowing that.
Everyone knows that very specific skills are needed for experiment
operations. These include the skills for designing experiments,
calibrating and operating instruments, measuring observational
parameters, presenting experimental results, and so on. In our
historical case, the major skills needed for operating the experiment
included those of setting up the arrangement for interferences by
reflection, producing a homogeneous light source, observing the
interference fringes, and measuring the changes of the fringes.
These experimental skills were employed in the original prismatic
interference experiment first conducted by Powell and the later
replications made by Potter and Airy, but none of them specified
these experimental skills clearly in the form of explicit descriptions

or instructions. One example is the skill of directly observing the image of the interference fringes with an eyeglass. Quite obviously the success of this observation technique relies on the relative positions of the eyeglass and the observer's eye. But nothing had been said on this issue in the whole debate. It seems that people should have known how to do so before they made the observation. Lack of explicit descriptions of this skill may not be accidental. It is quite possible that even these scientists themselves might not know how to describe this skill through a list of instructions, or that they might regard such a description as unnecessary, because they themselves did not master this skill in this way, or because they simply assumed this skill as part of the necessary expertise of every competent scientist.

In addition to the skills for experiment operations, there are more specific skills involved in the process of experiment replication. Many studies have indicated that a replication is not simply a matter of repeating an experiment identical with the original.[5] For those replications with the purpose of confirming the original experiment, an exactly identical experimental design usually provides a very limited support. In these cases, replications require redesigns of the experimental settings. For those replications with the purpose of disconfirming the original, though a completely identical setting is theoretically recommended, scientists usually need to make adjustments because of practical constraints. In these cases, replications should include a justification of the alterations. To make appropriate adjustments of experimental arrangements and to give convincing justifications of these adjustments require very sophisticated skills.

The involvement of experimental skills has a profound impact on the effectiveness of communication in the process of experiment replication. In the cases in which the experimental skills are totally tacit, namely, where there is no articulation of them at all in experimental reports, scientists have to figure out these skills by themselves, and misunderstandings occur easily. A good example in the prismatic interference experiment was the skills involved in producing a homogeneous light source with colored solution. At first Potter employed these skills in his experiment, but gave no description of them, making them completely tacit. When Airy tried to replicate Potter's experiment, he had to determine this tacit knowledge. From Potter's later paper we know that he used the solution of "iodine in hydriodic acid" rather than colored glasses to produce homogeneous light, because the former could generate purer and more intense light than the latter. But Airy did not know this. When he tried to identify the tacit knowledge behind Potter's light source on the basis of his own experience, which was quite probably limited to the uses of colored glasses, Airy concluded that Potter did not use homogenous light in his experiment at all! Clearly, this was a misunderstanding of Potter's experimental setting. But this was not Airy's fault, because Potter had left his experimental arrangement tacit. And this was not Potter's fault either, because he had every reason to assume that a first-rank researcher in optics like Airy would have the skills of producing homogeneous light that he did.

In those cases in which experimental skills are partially tacit, namely, where some descriptions of them have been provided, obstacles to effective communication still exist. The descriptions of experimental skills, even in the form of clearly written rules, have to be "reintegrated" into scientists' prior performances, as Polanyi has suggested. An example of this kind of complexity in the prismatic interference experiment was the skills involved in observing the interference fringes with an eyeglass. Potter did provide some details of his observational device. While the skills involved in determining the size and power of the eyeglass remained tacit, Potter clearly described the way he operated the device and the reference he used. Airy did not miss these explicit descriptions in his replication. But when he "reintegrated" this articulated knowledge with his prior practices, he inferred that in Potter's operation the eye must have left the eyeglass when withdrawing from the prism, which would bring about unreliable observations according to his experience. Airy designed an "instructive experiment" to show that keeping the eye focusing upon the eyeglass was necessary for observing the change of the interference fringes; later he adopted a new device that recorded a different experimental result. Hence, partial expressions of skills cannot eliminate the obstacles in communication.

For the sake of argument, we can even assume that in some cases skills can be articulated in a very explicit form such as a list of rules.[6] Even so, the difficulties in communication remain basically the same. The simplest example to illustrate this point is the case of following a rule of arithmetic such as "add a 2 and then another 2 and then another and so on." Although this rule has been fully articulated in language, several uncertainties remain when it comes to its applications, if we isolate this rule from our prior practices. For example, writing "82, 822, 8222, 82222," "28, 282, 2822, 22822," and even "8^2, 8^{22}, 8^{222}, 8^{2222}" may be said to be cases of following this rule. But in our daily life, we seldom apply this rule arbitrarily, because we integrate this rule with our prior practices, which provide a guideline for its applications. We obey this rule in a normal way because there are regular practices of following the rule in that way. We are trained to do so and through such training we firmly believe that what we do is simply the way it should be done.[7] Hence, integration with prior practice is crucial for every kind of skill-transference.

In short, the need for experimental skills imposes several constraints on the communication process in experiment replications.[8] First, these constraints stem from the fact that skills are seldom fully articulated in experimental reports. An effective skill-transference then may not be achieved merely through formal communication like writing and reading experimental reports. Informal communication, including private conversations and personal contacts, is crucial for the success of skill-transference. Whether an informal communication is possible however, in turn relies upon a series of contextual factors, especially the personal relationship between the scientists involved. Furthermore, constraints on the communication process in experiment replications also stem from the fact that integrations with prior practices are essential in the process of skill-transference. Consequently, contextual factors, such as the personal, social, and intellectual experience of the relevant scientists, must be involved in the process of skill-transference.

Therefore, if we take the important role of skills in experiment appraisal into account, we will understand why no agreement was reached on the experimental results of the prismatic interference experiment despite a series of replication attempts. In the appraisal of the prismatic interference experiment, the experiment produced a real physical effect, and both Potter and Airy tried to replicate each other's experiment "in the way prescribed." The disagreement between them, in the final analysis, resulted because the skills necessary for the experiment were not articulated fully in the experimental reports, and scientists had to integrate the knowledge presented in experimental reports with their prior practices in the process of skill-transference. Although the dispute between Potter and Airy could have been settled if the experiment had been performed directly in front of them, certain contextual factors, in particular, the differences in their intellectual and social status, finally led to a deadlock in the communication between them.

CONCLUSION

Popper presents an overly simplified picture of experiment appraisal. According to Popper, as long as an experiment produces a genuine physical effect, clear instructions of how to reproduce the experiment have been given, and scientists do replicate it in the way prescribed, they are likely to reach agreement on the acceptance or rejection of the result as physical evidence. If they do not agree with each other at the beginning, they can simply continue the replication process, and sooner and later resolve their differences. Popper is confident that an experiment displayed by real physical effects is bound to generate agreement among scientists through replication attempts. If not so, he claims, "scientific discovery would be reduced to absurdity, . . . [and] the soaring edifice of science would soon lie in ruins" (1959, 104). Following this line, Popper emphasizes the importance of giving clearly written instructions for further replications, which is necessary for achieving appropriate experiment replications. How to give accurate descriptions of experimental operations and results becomes a central issue in Popper's methodology for experiment appraisal. At the same time, factors such as scientists' personal and social characteristics are largely ignored.

However, in the appraisal of the prismatic interference experiment, we find that scientists reached no agreement, although the experiment produced genuine physical effects and the scientists did try to replicate the experiment in the way prescribed. As I have pointed out in the last section, what Popper has overlooked is the involvement of skills in the process of experiment appraisal. Experiment replications require an effective transfer of skills among relevant scientists, which usually are not fully articulated in linguistic descriptions. Therefore, appropriate experiment replications cannot be achieved solely by reading linguistic descriptions or instructions. Furthermore, the process of skill-transference is highly sensitive to contextual factors that can determine the effectiveness of communication essential for experiment appraisal. Because of these contextual factors, an experiment displayed by genuine physical effects may not always generate agreement among scientists in experiment appraisal.

This new understanding of experiment appraisal has several practical implications for the methodology of experiment appraisal, particularly regarding the methods for achieving successful experiment replications. First, we should not expect that we can fully understand an experiment only by reading the experimental report, no matter how accurate the report is. Some crucial skills involved in the experiment may not be articulated in the report. Secondly, in addition to formal communication involving the exchanges of linguistic descriptions of experimental procedures and results, informal communication is also necessary. Sometimes, an experiment can only be fully understood by directly witnessing the whole experimental process, or by personal contacts with the experimenters. Lastly, besides the details of experimental designs, operations, and results, the persons who conducted the experiment are also important for the process of experiment appraisal. Their characteristics, including personal and social features, can play an important role affecting the process of experiment replications.[8]

NOTES

1. This is Popper's "material requirement" for "basic statements," which are used to describe both observational and experimental results. This requirement is connected with the fact that, without helps of instruments, we can only reliably detect the displacement of macroscopic bodies. Experiments in fields such as microphysics and psychology, hence, need special instruments to convert microscopic or psychological effects into genuine physical effects, namely, the displacement of macroscopic bodies.

2. Later Potter received formal education at Cambridge, graduated in 1838 as a sixth Wrangler, and occupied a professorship of natural philosophy at University College, London from 1841 to 1865. See *Dictionary of National Biography,* Vol. 16, p.219.

3. Hamilton to Adare, (April 22, 1833), in Graves (1882, vol.2, 44).

4. This was the viewpoint of Hamilton. In the same letter to Adare, Hamilton wrote that "Airy is right, I think, to stop, for it was in danger of becoming too personal a matter."

5. For more discussions, see Franklin and Howson (1984), Collins (1984), Mulkay and Gilbert (1986).

6. To what extent skills can be articulated in language is controversial. For information about the relevant debates, see Sanders (1988, 5–6).

7. For more discussion on why practices are prior to the formulations and followings of rules, see Wittgenstein (1958, 75–88).

8. I am very grateful to Peter Achinstein, Peter Barker, Nathan Tierney, and Bill Angelett for many valuable comments on the drafts of this article.

REFERENCES

Airy, G. 1833a. "Remarks on Mr. Potter's experiment on interference," *Philosophical Magazine* 2:161–7.

———. 1833b. "Results of repetition of Mr. Potter's experiment of interposing a prism in the path of interfering light," *Philosophical Magazine* 2:451.

Franklin, A., and C. Howson. 1984. "Why do scientists prefer to vary their experiments?" *Studies in History and Philosophy of Science* 15:51–62.

———. 1988. "It probably is a valid experiment result: A Bayesian approach to the epistemology of experiment," *Studies in the History and Philosophy of Science* 19:419–27.

Graves, R. 1882. *The Life of Sir William Rowan Hamilton.* Dublin: Hodges, Figgis, & Co.

Hankins, T. 1980. *Sir William Rowan Hamilton.* Baltimore: The Johns Hopkins University Press.

Mulkay, M., and C. Gilbert, 1986. "Replication and mere replication," *Philosophy of Social Science* 16:21–37.

Polanyi, M. 1966. *The Tacit Dimension.* Garden City, NY: Doubleday.

———. 1969. *Knowing and Being. Essays by Michael Polanyi.* London: Routledge & Kegan.

Popper, K. 1959. *The Logic of Scientific Discovery.* London: Hutchinson.

Potter, R. 1833a. "On the modification of the interference of two pencils of homogeneous light produced by causing them to pass through a prism of glass, and on the importance of the phenomena which then take place in determining the velocity with which light traverses refracting substance," *Philosophical Magazine* 2:81–95.

———. 1833b. "A reply to the remarks of Professor Airy and Hamilton on the paper upon the interference of light after passing through a prism of glass," *Philosophical Magazine* 2:276–81.

———. 1833c. "Particulars of a series of experiments and calculations undertaken with a view to determine the velocity with which light traverses transparent media," *Philosophical Magazine* 3:333–42.

Powell, B. 1832a. "On experiments relative to the interference of light," *Philosophical Magazine* 11:1–7.

———. 1832b. "Further remarks on experiments relative to the interference of light," *Philosophical Magazine* 1:433–8.

Ryle, G. 1949. *The Concept of Mind.* Chicago: The University of Chicago Press.

Sanders, A. 1988. *Michael Polanyi's Post-critical Epistemology.* Amsterdam: Rodopi.

Wittgenstein, L. 1958. *Philosophical Investigation.* Oxford: Basil Blackwell.

4

STOCHASTIC ELECTRODYNAMICS AND COUNTERREVOLUTIONARY PHYSICS

Niall Shanks

Only after we have overthrown, finally vanquished, and expropriated the bourgeoisie of the whole world, and not only of one country, will wars become impossible. And from a scientific point of view it would be utterly wrong and utterly unrevolutionary for us to evade or gloss over the most important thing, namely . . . to crush the resistance of the bourgeoisie.

V. I. Lenin.

ABSTRACT

The central concern of this essay is the curious phenomenon of counterrevolutionary science. According to Kuhn, after a scientific revolution results in the instantiation of a new scientific orthodoxy, the practitioners of prerevolutionary science will fade away, their numbers being trimmed by old age and death. Counterrevolutionary science, however, is an attempt to account for the problem-phenomena which generated the crisis resulting in a scientific revolution, as well as phenomena uncovered by postrevolutionary investigators, from the standpoint of the concepts and analytical techniques characteristic of the prerevolutionary world view. The practitioners of counterrevolutionary science are typically too young to be mere holdovers from prerevolutionary days. The case study presented here concerns counterrevolutionary electrodynamics—attempts in the middle to late twentieth century to account for quantum electrodynamical phenomena from a classical standpoint.

INTRODUCTION: REVOLUTION AND COUNTERREVOLUTION IN PHYSICS

This essay is concerned with the curious phenomenon of counter-revolutionary science. More precisely, I will consider drawing some lessons for the history and methodology of science from a particular case study in counterrevolutionary physics—a branch of classical electrodynamical theory called stochastic electrodynamics (SED). This is a theory developed by a relatively small number of dedicated adherents (who have published a considerable number of papers on the subject, for the most part in quite respectable scientific jour-nals)—mainly over the last thirty years. Contemporary stochastic electrodynamicists are therefore too young to be mere holdovers from prerevolutionary physics, doomed to fade away as their num-bers are trimmed by old age and death. Rather, these theorists appear to belong to a small, though lively and vital, alternative scientific tradition. SED stands opposed to the dominant quantum electrodynamical orthodoxy, and is both classical and "prerevolu-tionary" in spirit.

In order to see why the phenomenon of counterrevolutionary science is interesting, it will be useful to consider some of Thomas Kuhn's views on the role and consequences of scientific revolutions and paradigmatic science [1970(a),(b)]. Since part of the aim of this essay is to introduce the reader to SED, some consideration will be given to classical electrodynamics (CED)—and in particular to the crisis-problems which led physicists to abandon CED in the early years of this century. Since the failure of CED was instrumental in prompting the quantum revolution, some of the main conceptual implications of this revolution will then be presented in order to provide a backdrop against which the development and character of SED can be discussed. Finally, consideration will be given to issues concerning the motivation for counterrevolutionary science and its implications for issues surrounding the unity of science.

KUHN ON SCIENTIFIC REVOLUTIONS

In the Kuhnian tradition in philosophy of science, it has become commonplace to present key events in the development of science using a revolutionary metaphor. From this standpoint, the history of science is viewed as a history of changing conceptual regimes. A paradigm is established, and there follows a period of problem-solving "normal science"—a period when the innovative concepts and practices embodied in the paradigm are applied to problems of interest. Needless to say, some problems resist treatment. At first this may not matter, but after repeated attempts to solve a problem from within the paradigm fail, the problem may take on a more sinister aspect. When it emerges that there are other seemingly untreatable problems, a sense of crisis may emerge within the scientific community.

This in turn may culminate in a period of revolutionary science—a turbulent period in the development of scientific ideas in which old, tried and trusted methods are abandoned and scientists consider employing new, and perhaps radically different concepts, methods and practices in order to find resolutions of the troubling problems. This in turn may result in the establishment of a new paradigm embodying concepts, methods and practices that are discontinuous with those found in the old regime. The discontinuity may be so serious as to affect the meanings of scientific terms, for as Kuhn has commented:

> In the transition from one theory to the next words change their meanings or conditions of applicability in subtle ways. Though most of the same signs are used before and after a revolution—e.g., force, mass element, compound, cell—the ways in which some of them attach to nature has somehow changed. Successive theories are thus, we say, incommensurable. [1970a, 266–67]

Kuhn adds, for example:

> . . . when teaching the development of Dalton's atomic theory, I point out that it implied a new view of chemical combination with the result that the line separating the referents of the terms "mixture" and "compound" shifted; alloys were compounds before Dalton, mixtures after. . . Part of learning to translate a language or a theory is learning to describe the world with which the language or theory functions. [1970a, 269–70]

The problems which prompted the revolution may receive their resolution in terms of concepts and a theoretical vocabulary that would seem quite alien to the prerevolutionary scientists.

After the revolution has established a new paradigm, science is *done* differently from the way it was done under the *ancien régime*— what counts as legitimate scientific activity may thus be transformed—and the remaining adherents to prerevolutionary ways are gradually marginalized. As Kuhn puts it:

Gradually the number of experiments, instruments, articles, and books based upon the paradigm will multiply. Still more men, convinced of the new view's fruitfulness, will adopt the new mode of practicing normal science, until at last only a few elderly hold-outs remain. . . Though the historian can always find men—Priest, for instance—who were unreasonable to resist for as long as they did, he will not find a point at which resistance becomes illogical or unscientific. At most he may wish to say that the man who continues to resist after his whole profession has been converted has *ipso facto* ceased to be a scientist. [1970b, 159]

Now this does not seem to be entirely correct. A close look at the history of physics in the twentieth century indicates that there has been a lively, albeit a minority interest in the concepts, methods and practices of prerevolutionary classical physics—and not just among elderly hold-outs. In short, there have been counterrevolutionary movements. And part of the concern of this essay is to arrive at an understanding of some of the conditions which must obtain in order for counterrevolutionary science to survive in the hands of theorists who cannot be properly dismissed as senile conservative reactionaries.[1]

THE *ANCIEN RÉGIME:* CLASSICAL ELECTRODYNAMICS

Electrodynamics can be thought of as the science of electric and magnetic fields and their mechanical causes and effects. Classical electrodynamical theory (CED) rests on two distinct groups of laws. First there are Newton's laws for particles:

1. Every particle continues in its state of rest or uniform motion in a straight line, except in so far as it is compelled by external forces to change that state.
2. The rate of change of the momentum of a body is proportional to the force acting and takes place in the direction of action of that force ($F = $ ma).
3. To every action there is an equal and opposite reaction.

Secondly, there are Maxwell's laws of electromagnetism. These equations define the electric and magnetic fields and express precisely their mutual relationships and the way they evolve with time.[2] Particles in CED behave in accord with Newton's second law under the Lorentz force:

4. $F = q[E + (v/c) \times B]$,

which describes the influence of the electric and magnetic fields E and B, respectively, on a moving charge q, traveling with velocity v and where c is the speed of light.

In addition to the laws of CED, care must be taken to specify "initial" or "boundary" conditions, for as Boyer, a leading stochastic electrodynamicist, points out:

> The interactions between particles and fields are accounted for by New-ton's laws of motion and by James Clark Maxwell's equations of electro-magnetism. In addition, certain boundary conditions must be specified if the theory is to make definite predictions. Maxwell's equations describe how an electromagnetic field changes from place to place and from moment to moment, but to calculate the actual value for the field one must know the initial or boundary values of the field, which provide a baseline for all subsequent changes. (1985, 73)

It is important that the chosen boundary conditions are such that all electromagnetic fields have moving charges as their sources. *In particular there are no source-free electromagnetic fields.* CED is deterministic, so for any system in its domain, once a complete state of the system is specified at some time t, the laws of CED will determine a unique state of the system for all other times. All physical processes described by the theory are not only determinis-tic, but continuous as well: particles have continuous trajectories and fields vary continuously from place to place.

Classical Electrodynamics: The Crisis-Problems

The failure of CED to address certain physical puzzles clearly in its domain led to its downfall and prompted the quantum revolution. In this connection, two puzzles deserve special attention: the prob-lem of blackbody radiation and the problem of atomic collapse.

First, the problem of blackbody radiation: this problem arose from a consideration of the continuous spectrum of electromagnetic radia-tion emitted by blackbodies. Blackbodies are objects that have the property of absorbing all the electromagnetic radiation incident upon them. Substances such as soot or coal may be viewed as approximations to ideal blackbodies. Experiments had shown that when a blackbody was heated to a given temperature, it would emit a maximum amount of energy at a particular frequency, the frequency of the maximum emission point increasing with increasing temperature. For frequencies higher than the maximum emission frequency, the emitted energy would rapidly tail off.

From the standpoint of theoretical analysis, a blackbody was considered to be a collection of oscillators in equilibrium with a field of electromagnetic radiation. Such an oscillator will absorb energy from the radiation field and radiate the energy back at a rate proportional to the equilibrium temperature T. The equilibrium condition is that the absorption and emission rates should be equal. In this case CED predicts the Rayleigh-Jeans law for the blackbody radiation energy spectrum.[3]

The Rayleigh-Jeans "ultraviolet collapse"—and hence the failure of CED—arises because the principle of equipartition of energy requires that each mode of oscillator vibration should be associated with the same amount of energy. Thus most of the energy from a blackbody should be given out at high frequencies. For a given temperature there will be no frequency at which a maximum amount of energy is emitted![4]

The problem of atomic collapse, is a little easier to grasp. Put simply, if, as Lord Rutherford had suggested, the electron in a (hydrogen) atom orbited the nucleus, in a manner reminiscent of a planet around its sun, then according to Maxwell's equations, since it was a moving charge, it would radiate electromagnetic energy and quickly spiral into the nucleus. If any two unresolved puzzles can generate a Kuhnian crisis, surely it was these two!

THE QUANTUM REVOLUTION AND ITS CONSEQUENCES

These crisis-problems ultimately received their resolution in the quantum theory. By the 1930s, the quantum revolution had brought forth many conceptual and ontological innovations, but what emerges is deeply troubling to classical eyes. Some of the reasons for this are as follows:

1. Quantum systems behave in an irreducibly probabilistic fashion—they are ontologically probabilistic, so the element of probability does not arise from *our* ignorance of precise initial conditions. The element of probability is, so to speak, *in the world.*

2. Quantum systems are indeterminate—"particles" *qua* quantum systems do not have simultaneous exact values for all observables associated with them by the quantum theory (a particle in a definite state of the position observable will not be in any definite

state of the momentum observable—and hence will literally take no value for momentum—and vice versa). Such systems cannot be viewed as billiard balls writ small. Of the quantum atom, Heisenberg wrote:

> All its qualities are inferential; no material properties can be directly attributed to it. That is to say, any picture of the atom that our imagination is able to invent is for that very reason defective. An understanding of the atomic world in that primary sensuous fashion. . . is impossible.[5]

Moreover, (pairs of) spacelike separated quantum systems can exist in states whereby there obtains between them significantly physical relationships which do not, and indeed cannot, supervene on the individual "intrinsic" properties of each system (the consequences of such phenomena for the individuation of individual quantum systems is not fully understood at this time).[6]

3. In influential versions of the orthodox interpretation due to von Neumann and Wigner, measurement interactions—changes of quantum state which are both discontinuous, non-unitary and inherently probabilistic, and hence categorically different from physical interactions analyzed in terms of the Schroedinger equation—occur as a result of the intervention of the "consciousness" of the observing subject.[7]

4. Quantum theory requires the use of nonclassical probability theory—so joint probability distributions for observables such as position and momentum, which are always well-defined in classical physics because they are thought of as random variables defined on a common phase space, are not in general well-defined in the quantum theory.[8]

5. Because of the algebraic peculiarities of the quantum theory, it has been argued that we must abandon classical logic (essentially a Boolean algebra defined over a field of two elements) in favor either of a three-valued logic, or perhaps a two-valued logic with non-boolean operators)—so "p or q" could be a true proposition, despite the fact that neither "p" was true nor "q" was true.[9]

6. Quantum "fields" are treated as entities with discontinuous, discrete features—in the case of quantum electrodynamics (QED) the "field quanta" are photons. Quantum fields have characteristics which are somewhat different from the continuous deterministic wave-fields of CED—in some elementary approaches to field quantization, the resulting field is equivalent to a "many-particle" Schroedinger equation.[10]

The counterrevolutionaries accept the existence of the quantum phenomena which prompted the quantum revolution, and they accept the phenomena uncovered by post-revolution problem-solvers. But they consider the conceptual price to be too heavy. Meanings of terms such as "particle," "wave," "field," and "process" seem to have shifted—but in counterrevolutionary eyes the shift has been one from meaning to incoherence. Moreover, counterrevolutionary physicists have been unwilling to accept at face value the changes in inferential patterns mandated by the adoption of nonclassical logic and nonclassical probability theory. The aim of counterrevolutionary physics is to save the quantum phenomena while preserving as much as possible of the ontology, epistemology, logic and semantics of the classical prerevolutionary world-view.

RENEGADE PHYSICS: STOCHASTIC ELECTRODYNAMICS

Stochastic electrodynamics is based on a determinate ontology of classical particles and waves—just like the entities in CED. These objects enjoy a deterministic dynamics. The techniques of statistical analysis are derived from classical probability theory. The logic of the theory is a two-valued (Boolean) logic. SED[11] is based on the same laws as CED, but on a nonstandard choice of boundary conditions.[12] In particular, the theory differs from CED in that it postulates the existence of a source-free field of classical electromagnetic radiation. As pointed out by Boyer:

> This theory assumes that the homogeneous solution to Maxwell's equations involves random classical radiation. . . The aspect of randomness in the theory involves the averaging over many microscopic but deterministic degrees of freedom. (1975, 796)

Thus stochastic electrodynamics is formulated to be consistent with a deterministic view of the world—the element of randomness arising in essentially the same way as it occurs in classical statistical mechanics.[13] The nonstandard choice of boundary conditions amounts to the view that in addition to electromagnetic radiation traceable back to moving charges, there is a source-independent universal background field of classical electromagnetic radiation. The stochastic electrodynamicists are not interested in the "first causes" of such electromagnetic radiation—any more than they are concerned with the "first causes" of matter. Rather, they wish to trace the consequences of its postulation.

The postulated random classical radiation field is to have the properties of homogeneity (so no position in space is preferred—for example, the distribution and density of the random radiation field will not vary from place to place), isotropy (so no direction in space is preferred), Lorentz invariance (so there are no preferred inertial frames), and temperature independence (so it is a "zero-point" field—unlike the electromagnetic radiation spoken of in CED, the random radiation field is not thermal in nature and persists at absolute zero). The random radiation field has an average energy of $1/2h\omega$ per normal mode of angular frequency ω (where h is Planck's constant)—a feature it shares with the vacuum field of QED![14]

The assumption that the zero-point field of SED has an average energy of $1/2h\omega$ per normal mode implies that h determines the magnitude of the field fluctuations. *It is at this point, and this point only, that Planck's constant enters SED, and it does so in a manner in no way suggestive of field quantization or quantization of action.* All processes in SED are both deterministic and continuous.[15]

Stochastic Electrodynamics and the Quantum Puzzles

The resulting electrodynamical theory has proven to be somewhat remarkable. Boyer (1969), (1975), (1980), (1983) and (1984b) has shown how SED can produce a classical, prerevolutionary account of the Planck radiation law—one devoid of all overtones of quantization. Adjusting the *boundary conditions* of CED, rather than abandoning the *laws and concepts* of CED, is evidently a strategy to be considered. This is significant, for it suggests that the empirically discovered blackbody spectrum did not *force* the quantum revolution in the sense of conceptually necessitating or mandating either the quantization of action or field quantization.[16] The "shape" or "form" of the blackbody radiation curve for blackbodies at given temperatures—hence the Planck law—may be determined by the evidence, but its interpretation (physical meaning or theoretical significance) is not.

What about the problem of atomic collapse? SED retains the planetary model of the atom. Boyer (1975, 800) has produced calculations on the basis of SED establishing the stability of the ground state of the hydrogen atom. He describes the heuristics of his approach as follows:

> The electrons moving about the nucleus are indeed radiating away energy according to classical electrodynamical calculations. However, a new element enters. The random zero-point radiation acts to produce random motions of the electrons, in effect transferring energy to the electrons by random classical electromagnetic forces. It is the balance between the energy loss by radiation and the energy pickup from zero-point radiation which must account for the stability of matter in random electrodynamics. (1975, 799–800)

Apparently, both of the crisis-problems which prompted the quantum revolution may be treated classically.

What conclusions should be drawn from the ability to extend prerevolutionary physics to blackbody radiation and the problem of atomic collapse? The main lesson is that we have concrete evidence here for the claim that theory is underdetermined by empirical data. Physics may, as a matter of fact, have evolved along the quantum route, but the crisis-problems did not necessitate this particular species of conceptual evolution. If SED explained nothing more than blackbody radiation and the problem of atomic collapse, it would be epistemologically significant for this reason alone.

What is of more interest is that SED has been applied to a wide range of quantum phenomena.[17] Of particular interest here is the account from within SED of van der Waals forces and the Casimir effect. The van der Waals force can be thought of as an attractive force existing between atoms and molecules.[18] The Casimir effect is a macroscopic effect generated by (attractive) van der Waals forces between (closely spaced) conducting plates. Reviewing this work, Milonni comments:

> In the calculations based on random electrodynamics . . . the physical interpretation is completely clear. Perhaps more remarkable than the fact that random electrodynamics predicts an interaction between "unexcited" atoms is the fact that for all atomic separations, and for either electrically or magnetically polarizable particles, it predicts an interaction of precisely the same form as in quantum electrodynamics. (1976, 76)

One should not underestimate this achievement, for the account of van der Waals forces and the Casimir effect from the standpoint of QED is a matter considered by many to be one of the crowning achievements of the theory. Commenting on the stochastic electrodynamical analysis of van der Waals forces and the Casimir effect, Boyer notes, ". . . the case of calculation within classical theory allowed complete evaluation of the forces at finite temperatures, something which has not been done in quantum theory" (1980, 56). It is clear from the literature that SED constitutes a living "alternative," counterrevolutionary research tradition, and should not be viewed as a mere curiosity for its ability to cover the historical crisis-problems.

Stochastic Electrodynamics: Implications and Problems

It would not be correct to convey the impression that SED has, as a matter of fact, "saved all the quantum phenomena." For triumphs notwithstanding, it is not a theory which may claim empirical equivalence with QED, that is, there are phenomena covered by QED which are not saved by SED. Nor perhaps, as a minority-interest, counterrevolutionary movement in physics, is it likely to. I shall have more to say about this in the next section. Some phenomena have so far resisted treatment within the framework of SED. Most notable in this regard is the absence of a treatment of discrete atomic spectra—though this is not perhaps surprising, since it is only recently that SED has become mature enough to be applied to problems relating to atomic structure. It is not known at this time, therefore, whether the problem of atomic spectra will be resolved by normal, "problem-solving" counterrevolutionary science, or whether it will fester and generate a crisis of its own.

Moreover SED is subject to the philosophical worries concerning the postulation of unobservable entities. The zero-point field is not directly observable—it is literally a *hidden variable*.[19] Boyer responds to worries in this regard by advocating the "method of hypothesis." Boyer claims that the postulation of the zero-point field simply stands or falls with the predictions derived from the theory (1975, 796). This was precisely the strategy employed by wave theorists in the nineteenth-century debate about the nature of light. These theorists were faced with a similar worry about the unobservability of the medium of light-wave propagation.[20]

I will turn now to consider a number of issues concerning the significance of SED, its relations to the other sciences, and the methodological lessons we can draw from its existence.

COUNTERREVOLUTION IN PERSPECTIVE

Perhaps the main historical lesson to be drawn from this case study of counterrevolutionary physics is that there has been less conceptual homogeneity after the quantum revolution than one might have expected on the basis of a Kuhnian analysis of the history of science. Like the state in Marxist-Leninist theory, the concepts and tenets of prerevolutionary science are supposed to wither away after the revolution. Yet this has not happened— classical ideas have proved surprisingly resilient and have had a life which transcends the biological lives of the holdouts from the *ancien régime.*

The Role of Counterrevolutionary Science

The idea of thriving alternative scientific traditions which save *significant* phenomena covered by the extant paradigm is inimical to the picture of paradigmatic science painted by Kuhn. This is not to say that the exponents of the quantum electrodynamical orthodoxy are more "liberal" than Kuhn's conceptual totalitarianism suggests. Indeed, the reverse seems to be true—for counterrevolutionary movements in the context of the quantum theory have been the subjects of vigorous attempts at suppression (as can be seen from the history of the "hidden variables" debate—see Jammer [1974, Ch. 7]). Yet these movements survived.

Why should this be so? Classical electrodynamics was abandoned not for its commitments to unobservable objects, but because it seemed it could no longer save the phenomena—and the quantum revolution was driven in part by empirical discoveries at the turn of the century. But what is it that drives the counterrevolutionary exponents of SED? Certainly, the theory is not motivated by any new empirical discoveries which have created a crisis for QED. Rather, the aim of SED is to produce an electrodynamical analysis which will save *significant* phenomena in the domain of QED. It is unlikely that theories as rich as QED and SED will ever be shown to be exactly empirically equivalent. Indeed there is no reasonable expectation of empirical equivalence, for SED has a relatively small group of active theorists associated with it, and the task is huge. The idea of saving significant phenomena is important, for when research traditions collide, the "battle" is not fought everywhere at once—rather there are particular (significant) phenomena where the interpretative struggles are especially intense. Blackbody radiation and the problem of atomic collapse are cases in point—as is the classical foray into the area of van der Waals forces and the Casimir effect. Significant phenomena are typically those phenomena presented by exponents of the postrevolutionary orthodoxy as phenomena which cannot be saved by the *ancien régime,* or they may be new phenomena uncovered and saved by postrevolutionary problem solvers, or both. Ideally, the aim of counterrevolutionary science is empirical equivalence with the postrevolutionary orthodoxy, but there will be little interest in such theories if they do not minimally save the significant phenomena.

It may emerge that there are phenomena in the domain of QED which cannot be saved, no matter how hard one tries, using the analytical techniques of SED. Then, perhaps, it will be "game-over." But SED would still have done useful work if this were to happen— for we could learn to differentiate, on the basis of such discoveries, which phenomena were nonclassical in nature and which were not. That blackbody radiation and the problem of atomic collapse are not uniquely quantum in nature is in itself a valuable discovery.

It seems that SED owes its existence to a perceived conceptual incoherence in the postrevolutionary orthodoxy. The very shifts in meaning and ontology which seem to many to be the exciting part of the Kuhnian perspective on science, can go too far and provoke cries of incoherence and nonsense. For example, quantum-talk of "wavicles"—ontological chameleons, sometimes wave-like and sometimes particle-like—or of "particle-systems," which have a definite value for momentum, but no value for position—or which can be at two places at once (for example, in the two-slit experiment)—may evoke such reactions.[21] And it remains a fact that some scientists see the aim of science as involving a coherent description of (microphysical) reality—and are consequently dismayed by the instrumentalistic evasiveness of their quantum colleagues who would dodge the conceptual problems of the theory through the slippery claim that the aim of the theory was to simply account for measurement results and their probabilities—reducing the aim of science to the accountants' desire simply to get the numbers right. It is important to emphasize these points about why counterrevolutionary movements exist. One of the standard objections to such movements from the proponents of the quantum orthodoxy is that these counterrevolutionary movements predict no *new* phenomena. Rather they just cover (significant) phenomena antecedently covered by the prevailing postrevolutionary orthodoxy. This sort of objection misses the point of counterrevolutionary science: for its motivation springs from a perception that the postrevolutionary interpretative orthodoxy is incoherent and in some sense irrational. As Arthur Fine once said of quantum mechanics, "It is, rather, the blackest of black box theories; a marvelous predictor but an incompetent explainer" (1982b, 740). The idea that we should place our faith in the first theory to come along and cover a given range of phenomena—what one might call "the first come, first believed" approach to epistemology—seems to confuse temporal priority with conceptual priority.

Counterrevolutionary science is motivated by a perceived interpretative inadequacy of the postrevolutionary orthodoxy. But what is the philosophical significance of counterrevolutionary science? I now turn to consider this issue.

Counterrevolution and Convergent Realism

The conceptual space in which counterrevolutionary science flourishes is created in part by the fact that neither the mathematical form of the laws of nature, nor the evidence supporting them, fixes the interpretation (theoretical significance or meaning) of the phenomena they cover. The discovery of the mathematical form of the Planck law may have been driven by empirical discoveries concerning the energy spectrum of blackbody radiation. But the interpretation of that law, either in terms of quantized oscillator action, or a quantized radiation field, was not so driven or determined by the evidence. SED shows that a purely continuous and deterministic wave interpretation is possible. I will call the existence of nonstandard interpretations of phenomena dealt with in the postrevolutionary orthodoxy a manifestation of the *interpretative disunity of science.*

A general point to be made here is that there is nothing privileged or unique about the preferred interpretations of phenomena provided by the postrevolutionary orthodoxy. That an interpretation of phenomena may be preferred in any given instance may have a variety of explanations—some, perhaps, psychological or sociological. Such interpretations are not fixed by the dictates of nature.

Counterrevolutionary science has some interesting implications for some influential arguments advanced in favor of scientific realism. These arguments have the effect of rendering the phenomenon of counterrevolutionary science mysterious. But the phenomenon is real—counterrevolutionary science is a legitimate fragment of the history of the development of modern science. So perhaps the arguments in favor of scientific realism will require critical examination.

One view of scientific realism holds that the success of modern science is a significant phenomenon that requires an explanation—it is presumably not just a miracle. In this context, scientific realism is viewed as an empirical hypothesis to explain the success of science. (See Putnam [1978] and Boyd [1984].) The hypothesis rests on two principles:

1. Theoretical terms in mature science typically have *referents* in the physical world.
2. The laws of a theory belonging to mature science are typically approximately *true*.

The scientific realist will thus reject those analyses of theoretical terms which deprive them of referents—for example, instrumentalism.[22]

The claim advanced by the scientific realist is that the concepts of "truth" and "reference" are to play a causal-explanatory role in an account of the success of science. The idea is that scientists behave as they do because they believe (1) and (2) are true. Their strategy works because (1) and (2) are true. This position may be called *convergent realism*.

To see what is going on here, one must bear in mind that the discussion of scientific realism takes place against the backdrop of the *convergence of mature scientific theories*. Essentially this amounts to the view that earlier mature theories are "approximately less true" accounts of the *same* subjects referred to in later such theories (which are converging on the Truth). Convergent realists have not said much about the nature of maturity as a property of theories, or the difference between mature and immature theories. I suspect that mature scientific theories are those that would be characterized by a Kuhnian as theories illustrative of paradigmatic science (for example, Newtonian and post-Newtonian physics, but not ancient physics). But the convergent realist rejects the Kuhnian view of science and must come up with his own terminology. It is clear below that convergent realists understand our current best physical theories to be exemplars of mature science, as well as their immediate ancestors. (One mark of a mature theory is that it has theoretical terms in common with those found in our current best science).

The convergent realist rejects Kuhn's picture of how scientific theories succeed one another. The convergent realist believes that when one mature theory is succeeded by another, the shared theoretical terms refer to the same objects and properties of reality. As Putnam puts it:

> Yet it is a fact that we can assign a referent to "gravitational field" in Newtonian theory *from the standpoint of* relativity theory (though not to "ether" or "phlogiston"); a referent to Mendel's "gene" from the standpoint of present day molecular biology; and a referent to Dalton's "atom" from the standpoint of quantum mechanics. These retrospective reference assignments depend on . . . the "principle of benefit of the doubt" or the "principle of charity", but not on *unreasonable* charity. Surely the "gene" discussed in molecular biology is the gene (or rather "factor") Mendel *intended* to talk about; it is certainly what he should have intended to talk about! (1978, 22)

So we may say that Rutherford "intended" to talk about the electron Bohr spoke of in his theory of stationary states. And that Bohr in turn "intended" to speak of the same electron that we now refer to from the standpoint of contemporary quantum electrodynamics. Putnam should not be interpreted as suggesting that Rutherford and Bohr (or Mendel) had some prescient insight into the development of science that was years in the future. Rather, the claim is that we can make good sense of the development and success of science by assuming that Rutherford, Bohr (and the contemporary quantum electrodynamicist) were referring to the same type of object when they used the theoretical term "electron." Or as Putnam writes:

> And . . . the principle of benefit of the doubt dictates that we should, in these circumstances, take Bohr to have been referring to what we call "electrons". We should just say we have a different theory of the *same* entities Bohr called "electrons" back then; his term did refer. (1978, 24)

But the existence of successful counterrevolutionary research traditions—those capable of saving the (significant) phenomena covered by the post-revolutionary orthodoxy—calls into question the *causal-explanatory* role to be played by the concept of "reference." According to Putnam questions of reference are to be settled from the standpoint of our current best scientific theories. From this vantage point, earlier mature theories are taken to be referring to the entities that we now refer to. Hence the convergent realist, unlike Kuhn, is committed to the possibility of cross-theoretic reference assignments. If theories T_1 and T_2 are mature theories, and T_2 is the successor to T_1, the convergent realist is committed to there being a common stock of theoretical terms (for example, "electron," "electromagnetic field"), or to there being theoretical terms in T_1 that are translations of theoretical terms in T_2 (in Putnam's example above, Mendel's "factor" = our "gene").

These shared terms are to "attach to nature" in essentially the same way. To this end Putnam appeals to the causal theory of reference. According to this theory, the theoretical term "electron" picks out the same objects in all possible worlds that contain electrons, *regardless of how different those possible electrons may be from the electrons found in the actual world.* This is what is meant by the claim that the term "electron" is a *rigid designator.*[23] This theory allows the convergent realist to claim that theoretical terms shared by two theories refer to the same objects regardless of how differently those theories *describe* the objects in question. (In one possible world there are electrons described by T_1, in another there are electrons described by T_2). In this way the convergent realist hopes to drive a wedge between the issue of what theoretical terms refer to, and the way a theory describes and characterizes the objects of reference. To put it bluntly, for the convergent realist T_1 and T_2 are different theories of the same objects—differing theoretical descriptions do not imply differences in the reference of theoretical terms. By contrast, for (antirealist) Kuhnians, T_1 and T_2 are different theories of different objects—differing theoretical descriptions imply ontological relativity (questions concerning the reference of theoretical terms only admit of theory-relative answers).

But what of counterrevolutionary physics? Counterrevolutionary physics is a real historical phenomenon. But the behavior of counterrevolutionary scientists is not something that is best explained by convergent realism. The counterrevolutionary scientist does not want to be caught dead describing and characterizing the denizens and processes of the microcosm after the fashion of the postrevolutionary orthodoxy. After all, counterrevolutionary science is motivated by a perceived incoherence in the preferred interpretations of the postrevolutionary orthodoxy. For such a theorist, the way objects and processes are *characterized and described* is very important.

Recall that the scientific realist is trying to explain the success of (mature) science. Part of the explanation of this success is supposed to hinge on the cross-theoretic constancy of reference of theoretical terms. Later mature theories are supposed to be better theories (because approximately truer) of the same objects spoken of in earlier mature theories. But such a view of scientific progress renders the phenomenon of counterrevolutionary science mysterious.

Counterrevolutionary science, which seeks empirical equivalence with the postrevolutionary orthodoxy, may be explained as a reaction to the way postrevolutionary science attempts to *characterize and describe* objects and processes referred to by theoretical terms. Whether or not theoretical terms "rigidly designate" is neither here nor there to the working counterrevolutionary scientist. Indeed, the counterrevolutionary scientist could consistently claim to be referring to (many of) the entities that the quantum theorist refers to—only he describes them very differently. To understand counterrevolutionary science it is important to focus on the issue of the theoretical interpretation, characterization and description of objects and processes—for these are the issues which "drive" counterrevolutionary science. The causal theory of reference which separates questions of the reference of theoretical terms from issues surrounding the description and characterization of the putative objects of reference will explain very little of the phenomenon in question. Such an approach to reference, whatever else may be said in its favor, seems ill-suited to causal explanations of the existence and apparent success of the phenomenon of counterrevolutionary physics discussed here.

Furthermore, the convergent realist maintains that the ancestor T_1 is to be *approximately less true* than T_2 (though both are "converging on the Truth"). But the counterrevolutionary scientist wishes to "save the phenomena" with his counterrevolutionary theory T_3. (T_3 will characterize and describe objects and processes from the standpoint of the conceptual scheme embodied in the ancestor theory T_1). But the theory T_3 is not supposed to be "approximately less true" than the postrevolutionary "successor" theory, T_2. In the *ideal* case, the theory T_3 will be empirically equivalent to T_2. And in counterrevolutionary eyes T_3 will be preferable for, in addition to being empirically indistinguishable from its successor, it will possess the "classical" theoretical virtues (the virtues that appear to be absent in the postrevolutionary scheme), and none of the theoretical vices (the conceptual "innovations" of the postrevolutionary orthodoxy that are so abhorrent to counterrevolutionary theoreticians). Counterrevolutionary science teaches us that sometimes theories are advanced and developed not because they are *truer* than current theories, but because they seem to offer a more satisfying *description* of *known* phenomena.

So in addition to the problems with the concept of "reference," the phenomenon of counterrevolutionary science poses difficulties for the alleged causal-explanatory powers of the concept of "truth." Thus there are events in the history of the development of mature sciences—counterrevolutionary reactions—that are hard to reconcile with the basic tenets of convergent scientific realism.

Counterrevolutionary Physics and the Disunity of Science

I will close by considering a special type of objection which has been raised against SED. The objection is the *intertheory objection*. The objection springs from the alleged *nomic unity of science*. Concerning this, Pierre Duhem has written, "Physical theory has to try to represent the whole group of natural laws by a single system all of whose parts are logically compatible with one another" (1982, 293). The objection goes like this: SED, considered in and of itself, may be more or less consistent and interesting. But, it is seriously inconsistent with other theories we believe to be true—in particular, general relativity. The zero-point radiation field, with an average energy of $1/2h\omega$ per normal mode, has a divergent total energy:

$$(1)\ E = \Sigma_\omega\ 1/2h\omega = \infty.$$

Since this amounts to infinite mass-energy, there should, on the basis of the general theory of relativity, be macroscopic cosmological effects—and these we do not see. So we must reject SED.

We should be careful not to overstate the force of this species of objection, for it is ironically a double-edged sword which can be used to slash the quantum orthodoxy too. For one thing, the zero-point field of SED is an analog of the vacuum field of QED—and here sauce for the goose is sauce for the gander—as is shown by the discussion of the vacuum field provided by Feynman and Hibbs (1965, 244–46).[24] Moreover, the intertheoretic objection can be directly applied to the quantum orthodoxy itself—for so far no mathematically consistent way has been found to combine quantum field theory and general relativity. (See Isham 1989, 82.) To date quantum gravity is a field of learning strewn with paradox and unwanted consequences.[25] In fact some problems here seem to have spawned interpretative disunity in the quantum camp itself.[26]

The force of the intertheoretic objection derives in part from the alleged nomic unity of science: the assumption that orthodox physics (quantum theory and relativity theory) is itself a unified collection of mutually consistent theories such that if counterrevolutionary physics conflicts with any part of this body of knowledge, then in some sense it conflicts with all of it—like the one soldier who is out of step with the entire regiment! But this is substantially false. For the orthodoxy itself is riven by serious consistency problems and is thus much more fragmented and compartmentalized—and hence more conceptually inhomogeneous—than its apologists would like to admit. (See Cartwright [1983], 13–20.)[27] The facts point to the nomic *disunity* of science and the nomic disunity of science creates a space for counterrevolutionary physics by undermining the force of the intertheory objection.

Counterrevolutionary science is a curious phenomenon. To ignore counterrevolutionary research traditions is to do violence to the history of science. Yet once the phenomenon is acknowledged, it is clear that some popular models of scientific change and scientific progress will have to be revised. The history of modern science is not one of successive unchallenged paradigms. Nor can it be straightforwardly rendered as a simple progression of theories, with each temporal successor converging ever more closely to the Truth.

NOTES

1. In this essay, I am only interested in counterrevolutionary movements in science insofar as they attempt to save the phenomena—and hence strive to be empirically equivalent conceptual alternatives to the prevailing postrevolutionary orthodoxy.

2. For a good discussion of the history and mathematics of Maxwell's equations, see Walker (1984). Readers interested in a particularly lucid introduction to electrodynamics which presupposes only a very limited background in mathematics should consult Adair (1987, 133).

3. *The Rayleigh-Jeans law:* $\rho(\omega) = \omega^2 kT/\pi^2 c^3$, give the blackbody radiation energy spectrum. Here ρ is an energy density function, ω is a normal mode of oscillator vibration, T denotes the absolute temperature, k is the Boltzmann constant and c is the speed of light.

4. The Planck law for blackbody radiation is as follows: $\rho(\omega) = \omega^2/\pi^2 c^3 [h\omega/ \exp^{h\omega/kT} - 1]$. Here h is Planck's constant. Between 1900 and 1911 Planck struggled to find an interpretation of his law, and only by small degrees did he come to see quantization of oscillator (emission) behavior as "inevitable." That is to say, Planck came to the view that oscillators did not emit radiation continuously, but rather in discrete packets or *quanta*. The relation between the energy of these quanta and the frequency of oscillation is $E = h\upsilon$. By contrast, Einstein had been happy to quantize the electromagnetic

field—introducing the idea of photons as "particles" of electromagnetic radiation—as early as 1905. A discrete electromagnetic field was, needless to say, a problem from the standpoint of the apparatus of CED.

5. Quoted in Mason (1962, 502). Given Heisenberg's pessimism concerning our attempts to picture quantum systems in a coherent fashion, small wonder many defenders of the quantum orthodoxy lapsed into an instrumentalist sulk. For a fuller discussion of the concepts of indeterminism and indeterminacy, consult Shanks (1991).

6. This feature of quantum theory has been termed "relational holism." See Teller in McMullin and Cushing (eds.), (1989).

7. For a useful examination of a number of issues surrounding the measurement problem in quantum mechanics, consult Cartwright (1983, Ch. 9).

8. For a fuller discussion of the issue of joint probability distributions, consult Fine (1982a).

9. For a good review of the history of quantum logical approaches, consult Jammer (1974, Ch. 8).

10. There are many deeply perplexing issues surrounding the interpretation of quantum field theory (QFT). Some of these receive a clear and accessible treatment by Redhead, in Brown and Harré (eds.) (1990).

11. Stochastic electrodynamics is sometimes referred to in the literature as random electrodynamics.

12. I am basing my exposition of SED largely on the work of T. H. Boyer. However, there are many other, extensively published investigators in this field. A reader wishing to know more should consult the extensive bibliographies in Boyer (1975) and (1980a); Milonni (1976); De la Pena and Cetto (1982) and Marshall and Santos (1987). All the above publications are worthy of study in their own right.

13. In classical statistical mechanics (CSM) it is granted that complex systems—for example, systems consisting of many particles—evolve deterministically in accord with Newton's laws in the sense that if the initial state of the system is known with absolute precision, a calculation based on Newton's laws will (in principle) enable us to calculate the precise physical state of the system at any other time. CSM was developed because standardly we are not in a position to know the precise initial states of complex systems. Moreover, the deterministic equations governing complex systems are typically themselves so complicated as to defy human computational capacities. CSM enables us to study complex systems on the

basis of partial knowledge of their physical states. The aim of the science is to make predictions as to how such systems will behave *on the average*. The element of randomness in the behavior of systems in the domain of CSM is epistemic—it arises from our ignorance of precise initial conditions.

14. The energy spectrum of the random radiation field is given by:

$$(1) \rho(\omega) = (\omega^2/\pi^2 c^3) 1/2 h\omega.$$

This is believed to be the only spectrum possible for a non-vanishing Lorentz invariant field. The one free parameter here is h. Defining h to be Planck's constant divided by 2π, permits the explanation of a range of quantum electrodynamical phenomena in terms of SED.

15. As Boyer points out, the zero-point radiation spectrum, "... is the unique spectrum of random classical electromagnetic radiation which is Lorentz invariant, isotropic in every inertial frame, invariant under adiabatic compression, and invariant under scattering by a dipole oscillator moving with arbitrary constant velocity" (1980, 52).

16. This would seem to support some of the conclusions that Quine drew in his [1961] against the empiricist dogma that evidence determines theory.

17. The successful applications of SED, as well as problems which have so far not received a satisfactory treatment, are reviewed in Boyer (1975), (1980), De la Pena and Cetto (1982), and Milonni (1976). In addition to blackbody radiation and the problem of atomic stability, SED has been successfully applied to a wide variety of oscillator systems. Sachindanandam (1983) has extended SED to cover the quantum phenomenon of intrinsic spin. Marshall and Santos (1987) have shown that stochastic optics (a branch of SED) is capable of covering a significant range of phenomena in the domain of quantum optics—with applications to the Aspect-type photon correlation experiments which are considered to be particularly telling against all hidden variables interpretations of quantum phenomena (of which SED is one). Moreover, in his (1984a) Boyer shows how the thermal effects of acceleration predicted within quantum field theory, are likewise predicted by SED. At the level of qualitative explanations, Boyer's work on SED (1975, 1980) has yielded an account of the two-slit experiment—one essentially similar in strategy to that offered by Bohm and Hiley (1982); the random disturbances of macroscopic measuring devices on microscopic systems; and a treatment of the Heisenberg relations suggestive of their

interpretation as statistical scatter relations in a manner reminiscent of Ballantine (1970). Finally, there exist numerous attempts in the SED literature to derive in classical terms, laws of motion equivalent to both the Schroedinger and Dirac equations—see Davies and Burkitt (1980) and Nelson (1966).

18. Van der Waals forces appear in the van der Waals gas equation, which represents ordinary gases in a more satisfactory manner than the ideal gas equation ($pv = RT$). The van der Waals gas equation is as follows:

$$(1) \; [p + (a/v^2)] \, (v - b) = RT,$$

where p = pressure, v = molar volume, T = absolute temperature, R = the gas constant, and a and b = positive constants. The term a/v^2 represents the van der Waals forces between molecules of the gas. Van der Waals forces result in an effect similar to a slight compression of the gas.

19. The zero-point field in QED is not directly observable either. Moreover, postrevolutionary physics has not shirked from postulating entities unobservable in principle, as anyone familiar with the doctrine of quark confinement (or the postulation of unobservable singularities lurking beyond event-horizons) is well aware. While there is an interesting philosophical debate about the ontological status of "unobservables" in physics, the fact remains that they play a vigorous role in the formulation of both pre- and postrevolutionary physics. The zero-point field in both QED and SED plays a role in causal explanations (of the Casimir effect, Lamb shift, and so on).

20. For details of the nineteenth-century debate, see Achinstein (1991).

21. Philosophers of science, insofar as they have been willing to clarify and elucidate the conceptual difficulties lurking within the quantum orthodoxy, may have unwittingly helped the counterrevolutionary process.

22. Instrumentalism is the view according to which scientific theories are merely devices to make predictions and "get the numbers right." The instrumentalist rejects the idea that science should aim to provide a coherent picture of the physical world. The instrumentalist attitude to theoretical terms—terms which refer to what is not directly observable (for example, "electron")—is that they are at best convenient fictions to oil the wheels of empirically adequate theories (theories which save the phenomena). At worst, such terms are merely meaningless "chicken tracks" on paper.

23. The reader interested in a lucid introduction to the causal theory of reference should consult Haack (1978), 58–60. A more advanced discussion of the causal theory of reference and the nature of theoretical terms may be found in Putnam (1977). The concept of "rigid designators" is due to Kripke (1972).

24. QED itself contains several different types of infinity—not just the one encountered in SED. In fact, QED has some highly controversial features. The method employed to avoid embarrassing infinities is called "renormalization." Of this analytical technique, Feynman makes the following comments: ". . . it is what I would call a dippy process! Having to resort to such hocus pocus has prevented us from proving that the theory of quantum electrodynamics is mathematically self-consistent. . . I suspect that renormalization is not mathematically legitimate" (1985, 128).

25. See also Penrose and Isham (eds.) (1986) for a collection of papers which explore these problems. As the editors put it, ". . . we wished to explore the possibility that the rules of quantum theory might need to be modified before a successful union with general relativity can be achieved". As for unwanted consequences, Ruger [1989] has explored some of the issues surrounding the peculiar implication of quantum gravity that whether or not certain particles exist depends on the state of motion of the observer—so the very existence of these particles is observer-dependent.

26. In some versions of quantum gravity reference is made to the wave-function of the entire universe. Since it seems implausible to suppose that there is a suitable *deus ex machina* whose consciousness is capable of reducing such a cosmic wave packet—and it seems grossly anthropocentric to suppose that *we* are responsible for such reductions—some interest has focussed on the "Many-Worlds" interpretation of quantum mechanics. On this interpretation—familiar to *aficionados* of science fiction—when a state evolves into a linear superposition of eigenstates of some observable—as happens in the paradox of Schroedinger's cat—the wave packet is never reduced, rather the entire universe "splits" into parallel universes, one universe for each component of the superposition. Schroedinger's cat becomes two cats, one alive and one dead.

27. Not all physicists think that the search for a unified framework is a good idea. There is the danger that unified theories would only apply to exotic phenomena, such as the big-bang, which cannot be tested experimentally—leading to what field theorist Howard Georgi has called "recreational mathematical theology." Georgi seems to believe that at different energy levels, different field

theories are applicable—and that we have no use for grand unified field theories—any more than a civil engineer has any use for the quantum theory of the atomic components of his or her girders. Georgi's views are discussed in Crease and Mann (1986, 414–17).

REFERENCES

Achinstein, P. 1991. *Particles and Waves.* New York: Oxford University Press.

Adair, R. K. 1987. *The Great Design* Oxford: Oxford University Press.

Ballentine, L. E. 1970. "The Statistical Interpretation of Quantum Mechanics," *Reviews of Modern Physics* 47: 358–81.

Bohm, D. J. and Hiley, B. 1982. "The DeBroglie Pilot Wave Theory and the Development of New Insights Arising Out Of It," *Foundations of Physics* 12: 1001–15.

Boyd, R. 1984. "The Current Status of Scientific Realism," in J. Leplin (ed.) *Scientific Realism.* Berkeley: University of California Press.

Boyer, T. H. 1969. "Derivation of the Blackbody Radiation Spectrum without Quantum Assumptions," *Physical Review* 182: 1374–1383.

Boyer, T. H. 1975, "Random Electrodynamics: the theory of classical electrodynamics with classical electromagnetic zero-point radiation," *Physical Review D* 11: 790–808.

Boyer, T. H. 1980. "A Brief Survey of Stochastic Electrodynamics," in A. O. Barut (ed.) *Foundations of Radiation Theory and Quantum Electrodynamics.* NY: Wiley.

Boyer, T. H. 1983. "Derivation of the Planck Radiation Spectrum as an Interpolation Formula in Classical Electrodynamics with Classical Electromagnetic Zero-point Radiation," *Physical Review D* 27: 2906–11.

Boyer, T. H. 1984a. "Thermal Effects of Acceleration for a Classical Dipole Oscillator in Classical Electromagnetic Zero-point Radiation," *Physical Review D* 29: 1089–95.

Boyer, T. H. 1984b. "Derivation of the Blackbody Radiation Spectrum from the Equivalence Principle in Classical Physics with Classical Zero-point Radiation," *Physical Review D* 29: 1096–1098.

Boyer, T. H. 1985. "The Classical Vacuum," *Scientific American* 253: 70–78.

Cartwright, N. D. 1983. *How the Laws of Physics Lie.* Oxford: Clarendon Press.

Crease, R. P. and Mann, C. C. 1986. *The Second Creation.* NY: Macmillan.

Cushing, J. T. and McMullin, E, (eds.) 1987. *Philosophical Consequences of Quantum Theory.* Indiana: University of Notre Dame Press.

Davies, B. and Burkitt, A. N. 1980. "On the Relationship between Quantum, Random and Semiclassical Electrodynamics," *Australasian Journal of Physics* 33: 671–84.

De la Pena, L. and Cetto, A. M. 1982. "Does Quantum Mechanics accept a Stochastic Support?" *Foundations of Physics* 12: 1017–37.

Duhem, P. 1982. *The Aim and Structure of Physical Theory.* NJ: Princeton University Press.

Feynman, R. P. and Hibbs, A. R. *Quantum Mechanics and Path Integrals.* NY: McGraw-Hill.

Feynman, R. P. *QED: The Strange Theory of Light and Matter.* NJ: Princeton University Press.

Fine, A. I. 1982a. "Joint Distributions, Quantum Correlations and Commuting Observables," *Journal of Mathematical Physics* 23: 1306–10.

Fine, A. I. 1982b, "Antinomies of Entanglement: The Puzzling Case of the Tangled Statistics," *Journal of Philosophy* 79: 733–47.

Haack, S. 1978. *Philosophy of Logics.* Cambridge: Cambridge University Press.

Isham, C. 1989. "Quantum Gravity," in Davies, P. (ed.) *The New Physics.* Cambridge: Cambridge University Press.

Jammer, M. 1974. *The Philosophy of Quantum Mechanics.* NY: Wiley.

Kripke, S. 1972. "Naming and Necessity," in Harman, G. and Davidson, D (eds.) *Semantics of Natural Language.* Dordrecht: Reidel.

Kuhn, T. S. 1970a. "Reflections on My Critics," in Lakatos, I. and Musgrave, A. (eds.) *Criticism and the Growth of Knowledge.* Cambridge: Cambridge University Press.

Kuhn, T. S. 1970b. *The Structure of Scientific Revolutions.* Chicago: University of Chicago Press.

Marshall, T. and Santos, E. 1988. "Stochastic Optics: A Reaffirmation of the Wave Nature of Light," *Foundations of Physics* 18: 185–223.

Mason, S. 1962. *A History of the Sciences.* NY: Collier.

Milonni, P. W. 1976. "Semiclassical and Quantum Electrodynamical Approaches in Non-Relativistic Radiation Theory," *Physics Reports* 25: 1–81.

Nelson, E. 1966. "Derivation of the Schroedinger Equation from Newtonian Mechanics," *Physical Review* 150: 1079–1085.

Penrose, R. and Isham, C. J. 1986. *Quantum Concepts in Space and Time*. Oxford: Clarendon Press.

Putnam, H. 1977. "Meaning and Reference," in Schwartz, S. (ed.) *Naming, Necessity and Natural Kinds*. Ithaca: Cornell University Press.

Putnam, H. 1978. *Meaning and the Moral Sciences*. London: Routledge and Kegan Paul.

Quine, W. V. 1961. "Two Dogmas of Empiricism," in *From a Logical Point of View*. NY: Harper and Row.

Redhead, M. 1990. "A Philosopher Looks at Quantum Field Theory," in Brown, H. and Harré, R. (eds.) *Philosophical Foundations of Quantum Field Theory*. Oxford: Clarendon Press.

Ruger, A. 1989. "Complementarity Meets General Relativity: A Study in Ontological Commitments and Theory Unification," *Synthese* 79:559–80.

Sachindanandam, S. 1983. "A Derivation of Intrinsic Spin One-Half from Stochastic Electrodynamics," *Physics Letters* 97A: 323–24.

Shanks, N. 1991. "Probabilistic Physics and the Metaphysics of Time," *South African Journal of Philosophy* 10: 37–43.

Teller, P. 1987. "Relativity, Relational Holism and the Bell Inequalities," in Cushing, J. T. and McMullin, E. (eds.) 1987. *Philosophical Consequences of Quantum Theory*. Indiana: University of Notre Dame Press.

Walker, G. B. 1984. "The Axioms Underlying Maxwell's Electromagnetic Equations," *American Journal of Physics* 53: 1169–72.

5

IS EVIDENCE HISTORICAL?

Laura J. Snyder

ABSTRACT

It is a common belief among philosophers of science that a theory has evidence in its favor only when the theory can correctly predict the occurrence of some new phenomenon. Recently, a historian of science has criticized this view, claiming instead that theories are evidentially supported only when they can explain some previously known phenomenon. Both positions entail the "historical thesis" of evidence, which holds that the time at which some information is known to be true relative to the invention of a hypothesis is relevant to whether (or how strongly) the information confirms the hypothesis. In this paper I examine the contrasting claims made in favor of predictions and explanations. I then argue that both positions are untenable, because in fact the historical thesis is false. This claim is clarified by an examination of the concept of evidence, and the introduction of a new theory of scientific evidence.

Suppose that some startling new phenomenon is observed. It is unlike anything which has been observed before, and for quite some time the scientific community cannot explain it. Then a new theory is proposed which is able to explain the phenomenon. Should we thereby consider this theory to have evidence in its favor? That is, should we count the fact that the theory provides an explanation for this phenomenon as evidence that the theory is true?

Many philosophers of science would say no. A common view is that explanations do *not* count as evidence for a theory's truth, only successful predictions of new phenomena do. If the new theory had predicted the occurrence of this startling phenomenon *before* it occurred, then (once it did occur) the theory would have evidence in its favor. Recently, however, this position requiring predictions for

evidence has been challenged by a historian of science who claims
that only explanations of previously known facts—and not predictions—can count as evidence for a theory.

Both sides in the dispute seem supported by some common intuitions. Predictions of new phenomena are extremely impressive, and
seem to endow scientists with a kind of mystical power to foretell the
future. At the same time, the view requiring that a theory explain
some known fact which other theories have been unable to explain is
also intuitively appealing; we may feel that a theory's success where
others have failed must indicate something about its truth. In this
paper I will put these intuitions to the test.

Before examining these opposing positions, I must clarify the
terms being used by them. "Prediction" does not necessarily refer to
a future event, as it does in our ordinary usage of the term, but to a
correct statement about some presently unknown phenomenon,
whether it is a past, present, or future event. Examples include
predicting that the extinction of the dinosaurs was caused by a
collision on earth of an enormous meteor; that there are six quarks;
that the ozone layer will be dangerously depleted in 100 years. Each
of these facts is not yet known to be true; once discovered to be true,
the prediction would be considered successful. On the other hand,
"explanation" as the term is used here has to do with facts already
known to be true. In this dispute, a theory is said to offer an
explanation when it can account for a known fact. For example,
Newton's theory of universal gravitation offered an explanation for
the previously known fact that the tides ebb and flow.

So the dispute over whether evidence must consist in explanations
or in predictions is actually concerned with the issue of when the
evidence is known to be true relative to the time that the theory is
proposed. If information e is known when theory T is proposed, and T
explains e, then e is "old evidence." If T entails e, but e is not known
when T is proposed, then e is "new evidence." In this paper I examine
both the claim that evidence must always consist in new evidence,
and the contrasting claim that evidence must always consist in old
evidence. I argue that both positions are untenable, because the time
at which e is known to be true relative to the theory's invention is not
relevant to its status as evidence. Hence I reject what has been called
the "historical thesis" of evidence, which holds that the time at which
e is known *is* relevant. Evidence can consist in explanations as well
as in predictions. This claim is clarified by an examination of the
concept of evidence, and the introduction of a new theory of scientific
evidence.

THE PREFERENCE FOR PREDICTION

In this section I will briefly outline three types of arguments for the view that only successful predictions can count as evidence for a theory.

The "No coincidence" Argument

In 1846, the planet Neptune was discovered after its existence and position had been predicted independently by two men, U. J. J. LeVerrier and John Couch Adams. Perturbations in the orbit of Uranus expected on Newtonian theory had led these men to conclude that there must be an unobserved body external to Uranus' orbit exerting an additional gravitational force on the planet. Using Newtonian theory, they were able to calculate mathematically the mass and orbit of this unseen planet. Acting upon LeVerrier's calculations, astronomers at the Berlin Observatory found the planet less than one degree from its expected location. From Newton's theory, it was possible to predict successfully the existence, position, and mass of a previously unexpected planet. This success was considered further evidence for Newton's theory of universal gravitation.

William Whewell, like other nineteenth-century intellectuals, praised the discovery of Neptune as a triumph of astronomy, which he termed the "Queen of the Sciences." Astronomy deserved this royal title because of its success at predicting phenomena: not only new instances of known phenomena, such as eclipses, but also novel and previously unexpected phenomena such as an eighth planet. Whewell argued—most notably in a published interchange with John Stuart Mill—that "to predict unknown facts found afterwards to be true is . . . a confirmation of a theory which in impressiveness and value goes beyond any explanation of known facts" (1857, 557).

Whewell's position (unlike the others I will discuss in this section) is that while explanations can count as some evidence for a theory, predictions are much stronger evidence. I will concentrate here on the argument for the evidential value of predictions, because Whewell's type of argument can be used to deny any evidential value to explanations at all. The reasoning behind this position is as follows. Whewell claims that the agreement of the prediction with what occurs (that is, the fact that the prediction turns out to be correct) is "nothing strange, if the theory be true, but quite unaccountable, if it be not" (1860, 273–4). For example, if Newtonian theory were not true, Whewell would argue, the fact that from the theory we could correctly predict the existence, location, and mass of Neptune would be bewildering, and indeed miraculous. He is using what Gilbert Harman (1965) calls the "inference to the best explanation" argument. That is, Whewell is claiming that the best explanation for the success of the prediction is that the theory which made the prediction is true. Further, he is claiming that because it is the "best" explanation, it is the "correct" one. Whewell is not claiming that there is no other logically possible explanation for the agreement between prediction and event; he admits that this agreement might be due to "mere chance" or "coincidence." But Whewell rejects this possible explanation as clearly inferior to the explanation that the theory is true.[1]

On the other hand, the mere fact that a theory can explain already known phenomena does not count as (such strong) evidence for a theory's truth. If a theory is invented for the express purpose of explaining some phenomenon, the fact that it does explain it is certainly no coincidence. The reasoning seems to be that, when a theory can explain some known phenomenon, the possible reasons for the theory's explanatory success are not merely either "coincidence" or "truth": another possibility is added—the "ingenuity" of the theory's inventor. Since the theorist knew what was needed to make the theory "successful" (that is, it needed to explain a certain known fact), it is not surprising that the theory should be able to do so. So the "best explanation" for the theory's success is no longer that the theory is true; rather, it is that the theory's inventor exercised much ingenuity in engineering a theory in order to explain the phenomenon. Whewell shares the suspicion with many contemporary writers that a theorist who knows the facts that need explaining is somehow inappropriately influenced by knowing what the theory has to do. Those who require predictions to rule out such "ingenuity" are suggesting that predictive successes—unlike successful explanations—are (as Whewell puts it) "beyond the stamp of ingenuity to counterfeit."[2]

The Falsification Argument

In distinguishing science from "pseudosciences" such as astrology, people often point to predictive capability. The power to make predictions, which are then confirmed, is considered by many to be the hallmark of scientific activity. This view, and the fact that it is so widespread among the general public, are due in great measure to the writings of the philosopher Karl Popper. According to Popper, science aims at "bold conjectures" about the world. Proper scientific method consists in inventing such conjectures, or theories, and then attempting to find evidence which proves them to be false. No evidence can ever prove a theory to be true, or even probably true; but a theory which has withstood the sincere attempt at being proven false (that is, which has not been "falsified") is considered to be "highly corroborated." The best that any evidence can do, then, is to corroborate a theory.

On this view, only predictions can count as evidence, because only predictions are "potential falsifiers" of a theory. If a theory explains a known fact, this information cannot potentially falsify the theory because it is already known that the fact explained is true. But if a theory predicts an unknown fact (and if this prediction is verifiable observationally) then it can function as a potential falsifier: we can attempt to observe whether this phenomenon does, in fact, occur. If it does not, then we have falsified the theory which predicted it. If the predicted phenomenon *does* occur, then the theory is corroborated— that is, tested and not yet falsified.[3]

For example, according to this view, the fact that Kepler's elliptical theory of Mars' orbit could explain Tycho Brahe's data is not evidence for the theory. Because Brahe's data were known, the fact that Kepler could explain them was not potentially falsifying for his theory. That is, the data did not provide a means for testing and possibly rejecting the theory. However, Kepler's general theory that the orbits of all the planets describe ellipses could be falsified. This theory entails predictions—for example, about as yet unobserved planets—that could be tested in order to falsify the theory. When Uranus was discovered in 1781, and Neptune in 1846, their orbits were found to describe ellipses; in this way, Kepler's theory received corroboration.

The Positive-Relevance Argument

Many recent accounts of evidence are quantitative, and use probability theory in order to explicate the concept. One type of quantitative account claims that for some piece of information to be evidence for a hypothesis, the information must raise the probability that the hypothesis is true. Some, following Rudolf Carnap, call this the "positive relevance" view of evidence; it is also referred to as the "increase in probability" view.

On this view, if h is the hypothesis and e is the putative evidence,

e is evidence for h if and only if
$$p(h/e) > p(h).$$

That is, e is evidence for h if and only if the probability of h given e is greater than the probability of h alone. The increase in probability view is generally taken to entail that only predictions can be evidence for a hypothesis. This can be demonstrated using Bayes' theorem, which follows from the probability calculus.

$$Bayes'\ theorem:\quad p(h/e) = \frac{p(h) \times p(e/h)}{p(e)}$$

It is claimed that, by definition, a known fact has a probability equal to one (it is "certain"). That is, $p(e)=1$. Now if $p(e)=1$, then, because h is consistent with e (because h explains e), $p(e/h)=1$. Hence, using Bayes' theorem,

$$p(h/e) = p(h).$$

That is, e does *not* increase the probability that h is true when e is a known fact that h explains. On this positive relevance view of evidence, then, the explanation of a known fact cannot be considered evidence for a theory.

To illustrate this view, let e be Brahe's observations and h be the hypothesis that the orbit of Mars describes an ellipse. Since Brahe's observations are a known fact, $p(e)=1$; and since h is consistent with e, $p(e/h)=1$. Using Bayes' theorem, we get the result that the probability that h is true given e is equal to the probability that h is true without e. Hence, e does not raise h's probability, and so it is not evidence for h.

A successful prediction, on the other hand, always counts as evidence according to this view. Because a prediction is not yet known to be true (it is not certain), $p(e) < 1$ when e is a prediction. And since h entails e (that is, h leads to the prediction that e), $p(e/h)=1$. Again using Bayes' theorem, we get

$$p(h/e) > p(h).$$

Kepler's hypothesis that the orbit of Mars describes an ellipse entails the prediction e' that tomorrow, Mars will be in a certain position. Since it is at present uncertain that e' is true (since it has not yet occurred), $p(e') < 1$. However, since h entails e', $p(e'/h)=1$. Therefore when e′ is found to be true, it will raise the probability that h is true, since $p(h/e′) > p(h)$.[4]

THE CASE FOR EXPLANATION

In a recent series of articles the historian of science Stephen Brush has argued against all views requiring that evidence be "new."[5] His claim is that a prediction cannot be "reliable" evidence for a theory; only explanations of known facts can count as evidence for scientific theories.[6] Brush illustrates his claim with several examples, including one drawn from general relativity theory. I will briefly examine this example, in order to explicate his position.

After Einstein devised his theory of general relativity, he found that it could offer an explanation of a previously known fact: it could account for the advance of Mercury's perihelion. This phenomenon (referred to as "Mercury's orbit") is the fact that Mercury is closest to the sun at different points of its orbit during successive revolutions around the sun. The presence of other planets causes the orbit of a planet to be perturbed—that is, the axis of the orbit's ellipse rotates slowly relative to the fixed stars. In the case of Mercury, the observed perturbation is larger than that which is calculated by Newtonian theory. Einstein was able to construct an elaborate derivation showing that general relativity theory entailed this discrepancy of 43″ of arc per century between the theoretical (Newtonian) value and the observed position. Einstein showed that if the initial orbit of Mercury is calculated using general relativity theory, and the value thus obtained is added to the Newtonian value of perturbation caused by other planets, this total exactly conforms to the observed result. Hence his new theory was able to explain a known fact previously considered inexplicable.

Einstein's theory also entailed two predictions of novel phenomena. One of these was the gravitational reshift of spectral lines (that is, the prediction that the frequencies of celestial spectral lines should be shifted to measurably lower frequencies compared to those of terrestrial spectral lines). This prediction was not accurately confirmed experimentally until 1960. The other, which was confirmed during Einstein's lifetime, was the phenomenon of light-bending. Einstein's theory entailed the prediction that light passing a massive body such as the sun on its way to Earth would be attracted by the body's gravitational force. As a result, light from a distant star should be deflected as it passes the sun on its way to the Earth. In 1915 Einstein made his final calculation of the degree of deflection which should occur. This result could be observed by comparing the relative positions of a star when the sun is near the path of its light and when the sun is not. The first position would be visible only during a solar eclipse, when the sun's own light would not be blocking the light from the nearby star. Einstein's prediction was confirmed during the solar eclipse of May 1919.

Brush claims that, contrary to what proponents of the predictivist view would argue, this prediction of a quite unexpected phenomenon was not better evidence than the explanation of Mercury's orbit; rather, the explanation was better evidence for the truth of general relativity theory. His reasoning is that

> a successful explanation of a fact that other theories have already failed to explain satisfactorily (for example, the Mercury perihelion) is more convincing than the prediction of a new fact, *at least until competing theories have had a chance* (and failed) to explain it. (1989, 1127, my emphasis)

By the time Einstein gave his explanation for Mercury's orbit, other theories had tried and failed to explain it. However, since light-bending was a newly discovered phenomenon, other competing theories had not yet had the opportunity to explain it. Besides claiming that the explanation was "more convincing," Brush makes the stronger, noncomparative claim that a successful prediction cannot become good or "reliable" evidence for a theory until other theories have had their shot at explaining the new phenomenon. (I call this the "equal opportunity" requirement of Brush's view.) He asserts that "light-bending could not become reliable evidence for Einstein's theory until those alternatives failed" (1989, 1127). Hence, Brush claims "it was only ten years after the initial report of light-bending" that it could be "plausibly asserted" that the phenomenon was evidence for Einstein's theory (1989, 1126).[7]

THE CONCEPT OF EVIDENCE

All of the views which claim either that evidence must consist in predictions of unknown phenomena or that it must consist in explanations of known facts agree with what has been called the "historical thesis."[8] This thesis holds that, on any plausible theory of evidence, the time at which e is known relative to the invention of h is relevant to whether e is evidence for h. In the case of the predictivist positions examined, e is evidence for h only if e was not known to be true at the time that h was proposed, but was discovered to obtain sometime thereafter. Brush's view also entails the historical thesis, by claiming that for e to be evidence for h, e *must* have been known to be true before h was proposed. Other views which conform to the general historical thesis are those which claim that the time at which e is known to be true is relevant to how *strongly* e confirms h: for example, Whewell's view that predictions are stronger evidence than explanations. In the remainder of this paper I will argue that the historical thesis is incompatible with any theory of evidence that captures how the concept is used in science. Hence both positions in the explanation versus prediction dispute are untenable.

What *is* evidence? That is, what is it to claim that some information e is evidence for, or confirms, a hypothesis h? Roughly speaking, it is to claim that e gives a reason for believing in the truth of h.[9] When we say that the spots on John's face are evidence that he has the measles, we mean that the presence of the spots gives a reason to believe that John has the measles. Note that evidence need not be *conclusive*.[10] There can be a reason to believe that John has the measles without it being true that he is so afflicted.

Reasons for belief can be either personal or impersonal. Personal reasons for belief are relativized to a particular person; that is, some information e might be person P's reason for believing a hypothesis. There are two ways in which a reason for belief can be personal. The first is factual: e is person P's *factual* reason for believing h only if P in fact believes h for the reason that e. There is also a normative way in which a reason for belief can be personal: e is person P's *normative* reason for believing h only if, given P's knowledge and beliefs, e gives a reason for P to believe h, whether or not P does believe h for the reason that e. Another way of expressing this normative relativization is that person P *should* believe h for the reason that e, even if, in fact, P does not believe h at all or believes h for another reason.

A reason for belief that is impersonal is one which is not relativized to any particular person. That is, if e is an *impersonal* reason to believe h, then it provides a reason for anyone to believe h, whether or not any particular person does, in fact, believe h for the reason that e, or whether any person should, given his other beliefs and knowledge, believe h for the reason that e. For example, the fact that the water in the puddle outside is frozen solid (e) is an impersonal reason for anyone to believe that the outside temperature is zero degrees Celsius or below (h), even if a particular person does not (and should not, given his lack of knowledge of the Celsius scale) believe h because of e.

In science, evidence is used in the *impersonal* sense. When a scientist publishes experimental results, claiming that these constitute evidence for a certain theory, she is not claiming merely that these results constitute her personal (factual or normative) reason for believing the theory. Rather, the scientist is claiming that the results constitute a reason—a reason for anyone—to believe the theory. Of course, others may disagree with this assessment; but in such cases the disagreement is over whether some information truly is a reason for everyone to believe h, or whether it is reason enough.

This impersonal use of evidence is seen in several professional activities performed by scientists. First, scientists engage in public debate over theory-choice. That is, they argue amongst themselves over which theory of a group of competitors is the most reasonable theory to believe. For example, some medical researchers are currently arguing over whether there is more reason to believe that HIV infection *causes* AIDS, or that it is not a cause but merely a coexisting condition.[11] Scientists generally invoke evidence in these debates (for example, the evidence that a small number of patients with AIDS do not seem to be infected with HIV). The use of evidence in debating over theory-choice demonstrates that scientists do not consider evidence to be a personal concept. If scientific evidence were personal in the factual sense, then there would be no basis for *debate* at all; scientists S1 and S2 could merely *report* the hypotheses they each believe and the reasons for which they believe them.

It might be argued, against my claim, that argument over theory-choice takes place at the level of normative personal reasons for belief: that is, that when a scientist argues that e is a reason to believe h he means that other scientists who share his beliefs and knowledge should believe h because of e. While scientists may make this kind of argument, by so arguing they are not utilizing the normative personal concept of evidence. In such argument an appeal is made to beliefs and knowledge assumed to be shared by a group, not to the specific beliefs and knowledge of a particular individual. Moreover, a scientist invoking evidence in theory-choice debate is claiming *more* than just "those scientists who share my knowledge and beliefs ought to believe h because of e." The scientist is claiming, rather, that e is a reason for *anyone* to believe h, even those who lack his knowledge and beliefs. Thus, for example, the medical researchers arguing for the hypothesis that HIV is not the cause of AIDS are claiming that their evidence is a reason for anyone—other medical researchers, doctors, politicians, the general public—to believe h, no matter what other beliefs they might have. This is clearly a use of evidence as an impersonal concept. Of course, there is not always enough evidence known to choose between theories on the basis of the evidence; but my claim is that on any plausible theory of evidence, it must be possible in principle for such a choice to be made.[12]

Another professional activity performed by scientists is to evaluate, and occasionally to reject, the evidential claims of other scientists. This activity also exemplifies the impersonal character of evidence in science. For example, in recent years the scientific community has judged that the experiment which was claimed to provide evidence for cold fusion does not provide evidence that cold fusion occurred. Note that this judgment is not about whether the experimenters Fleischmann and Pons had a personal reason (either factual or normative) to believe in cold fusion because of their experiment. Rather, scientists have concluded that the reported experimental data are not, in fact, reason for anyone to believe that cold fusion occurred. Fleischmann and Pons are therefore judged to have been incorrect in their claim that there is an impersonal reason to believe in cold fusion.

Any plausible theory of evidence (that is, one which captures how the concept is used in science) must account for this impersonal use of evidence in science. Various philosophical explications of the concept of evidence have been proposed. These are either objective or nonobjective. A general condition for objective concepts of evidence is as follows:

> *Objective Concept of Evidence:* whether e is evidence for h does not depend upon anyone's beliefs or knowledge about e, h, or anything else. Hence if some e is evidence for h, it is so regardless of what any person knows or believes.

Particular objective theories of evidence are given by Carnap, Hempel, Glymour, and Achinstein, among others.[13] All satisfy this general objective condition. For example, Hempel's inductive account of evidence claims that a universal hypothesis is confirmed by its positive instances; hence the law "all crows are black" is confirmed by the existence of a black crow. This is so whether any person knows or believes that the crow exists, and even if no person knows or believes that a black crow is evidence for the hypothesis "all crows are black." Glymour's "bootstrapping" model of confirmation builds upon this positive-instance account, bringing in the idea that parts of a theory can be used to confirm other parts. In

Glymour's account, e confirms h (relative to some theory T) regardless of whether anyone knows or believes that e is true or that e is evidence for h. Achinstein's view incorporates the following conditions: e must be true, h must be highly probable given e, and it must be highly probable that there is an explanatory connection between h and e.[14] Again, this view does not require that anyone knows or believes that e or h is true, or that any of these conditions obtain.

Objective theories of evidence have been preferred by many philosophers of science in part, I think, because any theory which satisfies this objective condition is able to capture the impersonal sense of evidence in science. (Note, however, that I am not claiming that *only* objective theories are impersonal; below I discuss a nonobjective impersonal theory of evidence.) Objective theories are not relativized to any person; when some information is deemed evidence for a hypothesis it is considered to be evidence for h for anyone. If the conditions of a particular objective theory are met, e is evidence for everyone. Conversely, if these conditions are not met, then e is not evidence for h for anyone. Thus objective theories can account for the fact that scientists invoke evidence as an impersonal concept in debating over theory-choice and in evaluating evidential claims.

Writers on the historical thesis have failed to realize that the historical thesis is incompatible with any theory of evidence that is objective. On any objective view, the *time* at which e is known relative to the invention of h cannot be relevant to whether e is evidence for h. As we have seen, the objective condition for evidence explicitly denies all relevance to what any person knows or believes about e, h, or anything else. So, for instance, if e is evidence for h, on an objective view, it is evidence for h whether or not any person knows or believes that e is true. On such views, if John has spots of a certain kind, then there is a reason to believe that he has measles, even if no one has seen these spots—that is, even if no person knows that e is true. Indeed on objective views of evidence, even if no person *ever* knows about the spots, they are still evidence that John has the measles. Clearly if e does not need to be known in order to confirm h, then it makes no sense to require that it must be known either before or after h is invented, in order to confirm it.

So if we want to accept the historical thesis, as many do, we must consider whether it is compatible with any *nonobjective* view of evidence, and whether any such view can capture the impersonal feature of scientific evidence. Defining "nonobjective" evidence in opposition to objective evidence, we get the following general condition:

Nonobjective Concept of Evidence: whether e is evidence for h at time t *does* depend on what someone (either a particular person or a particular group) knows or believes about e, h, and/or something else at t.

At first glance this approach looks promising in terms of the historical thesis, since nonobjective views consider what is known when the evidential claim is made to be relevant in determining whether e is evidence for h.

Let us first examine the most well-known nonobjective theory of evidence, "subjective Bayesianism." The subjective Bayesian view of evidence is a quantitative one, defining evidence in terms of probability (that is, as satisfying the standard axioms of the mathematical probability calculus). The subjective Bayesians are proponents of the positive-relevance view discussed in the first section of this paper; that is, they consider that evidence must raise the probability of a hypothesis. However, unlike objective probabilists, they define probability in subjective terms. According to them, the probability of h given e for a person P is defined as P's degree of belief in h given the information expressed by e. A person's degrees of belief are characterized in terms of his attitudes towards real or potential bets on the truth of h. Subjective Bayesians demonstrate that one's set of degrees of belief in all hypotheses must satisfy the probability calculus in order to avoid the situation that one's betting behavior would necessarily result in loss. If one's set of degrees of belief satisfies the probability calculus, then subjective Bayesians speak of this set as *coherent.* Hence on this view, evidence can be defined in the following way: e is evidence for h for person P if and only if (i) e increases P's degree of belief in h, and (ii) the degree of belief in h given e is part of a probabilistically coherent set of beliefs that P has.[15]

This nonobjective view of evidence is readily compatible with the historical thesis. In fact, prominent critics of subjective Bayesianism such as Glymour have rejected the view because of its supposed entailment of the predictivist side of the historical thesis.[16] Recall that in the first section the positive-relevance view of prediction was discussed. There I demonstrated how, using Bayes' theorem, it was claimed that only predictions of unknown facts could raise the probability of a hypothesis. The subjective Bayesian view also requires an increase in probability. Hence the subjective Bayesian is taken to require predictions for evidence.[17]

Although subjective Bayesianism seems to entail the historical thesis, this theory does not capture the nature of evidence as it is used in science. In fact, the subjective Bayesian view denies that there is an impersonal concept of evidence; it makes evidence only a personal matter. Relative to their different sets of degrees of belief, two people can (and often *must,* to satisfy the coherence condition) have vastly different degrees of belief in a particular hypothesis given the same evidence.[18] Indeed, the fact that John has spots of a certain kind on his face might be evidence for two contrary hypotheses. For John's doctor, the spots might be a reason to believe the hypothesis that John has the measles (and was not cursed by a witch). But at the same time, for John's superstitious grandfather the spots might be evidence for the contrary hypothesis that John has been cursed by a witch. Because the same evidence can equally well support two conflicting hypotheses, relative to the differing sets of beliefs two people may possess, the subjective Bayesian view does not provide a basis for a scientific community to debate over theory-choice. On the basis of e, the doctor will choose one theory, the grandfather another, and neither will worry about the disagreement between them.

To a limited extent, the subjective Bayesian view does allow for the evaluation of evidential claims made by scientists. If scientist P claims that e is his evidence for h, but P's degree of belief in h makes his complete set of beliefs incoherent, then we can reject P's claim that e is his evidence for h. If the doctor claims that e is evidence that John has been cursed by a witch, we can reject this evidential claim as incoherent with the doctor's complete set of beliefs (assuming that it *is* incoherent with her beliefs). However, since such evaluation is relative to a particular person's set of degrees of belief, we cannot reject any evidential claim if it is consistent with the person's set of degrees of belief. So, for example, we cannot reject the grandfather's claim that John's spots are evidence that he has been cursed by a witch, as long as this belief is part of a probabilistically coherent set of beliefs held by the grandfather. On this view, scientists would not be justified in rejecting the evidential claims regarding cold fusion made by Fleischmann and Pons, if these claims are coherent with the experimenters' personal sets of beliefs.

Accordingly, subjective Bayesianism fails to capture the impersonal nature of evidence as the concept is used by scientists. However, I am not claiming that only objective notions of evidence can do so. On the contrary, there is a nonobjective view which captures this impersonal nature of scientific evidence, and which has not been discussed by philosophers of science. This is a more plausible nonobjective view than subjective Bayesianism. The concept of evidence it proposes is nonobjective because it depends on people's knowledge or beliefs; yet unlike the subjective Bayesian view, it is not relativized to the knowledge and beliefs of any particular person. Instead, it is relativized to some "total current body of scientific knowledge." On this view, evidence for a hypothesis h is a reason for any person to believe h at time t, given some body of knowledge known at t. Thus, for example, spots of a certain kind are evidence for measles at the present time, because (given our current medical knowledge), the presence of these spots on a patient's body is a reason for anyone to believe that the patient has measles.

By a "total body of knowledge" I mean what Newton means by "phenomena," as the term is used in the *Principia*.[19] Newton suggests by his examples that the "phenomena" consist in facts that have been established by observation. Further, these facts are presumed to be noncontroversial; that is, the "phenomena" consist in facts that are agreed upon by scientists who are aware of the observations, and which would be agreed upon by other scientists once they were made aware of them. Note, then, that not every person needs to know or believe these facts. Nonscientific laypeople may well not know all or even most of the facts constituting the "total body of knowledge." Even among scientists, there will be many who would not realize the truth of certain facts even if they were made aware of the relevant observations; for example, a medical researcher may not be in a position to agree or disagree on certain facts of physical astronomy. Here we can appeal, somewhat roughly, to the notion of "scientific experts" who establish the contents of the current body of scientific knowledge. Medical researchers establish the body of medical knowledge, while astronomers and physicists establish knowledge about celestial bodies and their motions. The important aspect of this notion is that the "body of knowledge" consists in noncontroversial facts accepted by the scientific experts (a weaker version would include beliefs which are generally thought to be facts by the experts but might later turn out to be false).

When there is disagreement among the experts in a particular field, as there is now over the causal relation between HIV and AIDS, the disputed information is not part of this body of accepted knowledge. Hence this is a *temporal* concept; what is considered accepted knowledge changes over time. For example, today it is accepted by the experts that the moon has mountains and valleys on its surface. But before Galileo turned the telescope to the heavens and observed the moon, the experts believed that the moon's surface was smooth. Even when Galileo observed that the moon had an irregular surface, this fact was controversial for many years, until it was accepted that the telescope did indeed accurately represent celestial bodies. Until the scientific experts agreed, this fact was not part of what I am calling the "total body of knowledge."[20]

This nonobjective concept of evidence satisfies the following condition:

"Expert-relative" Concept: e is evidence for h at time t only if there is some body of information b that is known by the "experts" at t and, given b, e is a reason to believe h.[21]

This nonobjective concept of evidence, unlike the subjective Bayesian one, is able to account for the impersonal nature of evidence at any given time. On this view, evidence is not relativized to the knowledge and beliefs of any particular person; so if some e is determined to be evidence for h at time t, it is evidence for h (at t) for anyone. Unlike the objective views, this theory makes evidence a contextual matter, dependent upon the historical and epistemic circumstances surrounding the evidential claim. I will illustrate these aspects of the expert-relative concept of evidence by taking an intuitive (if somewhat gruesome) case. Suppose I discover a headless body with its severed head lying several feet away. The state of these body parts (e) constitutes very good reason for any person to believe the hypothesis (h) "this person is dead," relative to all current medical knowledge (b). As opposed to the subjective Bayesian view of evidence, e is evidence for h for everyone, even for those people who do not know or believe this medical knowledge. And on this view, in contrast to the objective views, e might not always have been evidence for h. The first time a decapitated body was observed by a primitive society, it would not have been a reason to believe that the person was dead, if the "total current body of knowledge" did not yet include the information that humans cannot live without their heads attached to their bodies.

This expert-relative notion of evidence allows for the invoking of evidence as an impersonal concept in debates over theory-choice. In this example, e (John's decapitated body) provides a basis for debating over which hypothesis is more reasonable to believe at the present time: h1 (John is dead) or h2 (John is not dead). Given our current knowledge, e is a reason, at this time, for *anyone* to believe h1 rather than h2. Further, on this nonobjective view it is possible to evaluate and reject evidential claims. And unlike on the subjective Bayesian view, these evaluations are not relative to any *particular* person's background beliefs or knowledge. On this view, e is evidence for h even if e is not my personal reason for believing in h (in either the factual or normative sense). For example, if I do not know that medical knowledge dictates that headless human bodies cannot be alive, or if I disagree with the experts on this, I might not believe h on the basis of e. Yet regardless of what I believe, e would still be an excellent reason now (given b) to believe that the person is dead. My claim that e is not evidence for h can—and should—be rejected.

On the expert-relative view, evidential claims must be evaluated contextually, relative to the contents of b at the time the claim is made. For this reason, the expert-relative notion of evidence can be useful in studying the history of science, particularly in cases where evidential claims are sound, relative to what was known by the experts at the time they were made, but unsound given what we know today. For example, in the second century A.D. Ptolemy argued that the Earth does not rotate about its axis (h). He claimed that his evidence for h was that when a stone is tossed up vertically, it is observed to land straight down in the same position from which it was thrown (e). Ptolemy asserted that e was a reason to believe h because, if the Earth were rotating eastward, then the stone should be observed to land west of its starting point (because the Earth would have rotated slightly during the time the stone was in the air). Relative to his knowledge and beliefs e increased his subjective degree of belief in h (he already had a high prior degree of belief in h); further, believing in h to the degree he did was coherent with Ptolemy's complete set of beliefs. So, with the subjective Bayesians, we can say that e was Ptolemy's *personal* reason for believing h. But we can make a stronger evidential claim than this. It is also the case that, relative to the total body of knowledge of his time (which did not include knowledge of inertial motion), e was a reason for *anyone* to believe h. Relative to the contents of b at time t, e was an *impersonal* reason to believe h. So we can use this expert-relative notion of evidence to evaluate Ptolemy's evidential claim in the impersonal sense, even though, relative to our current knowledge, e is not an impersonal reason to believe h.[22]

This nonobjective concept of evidence, then, is able to capture the impersonal nature of evidence at a given time in science. Yet, like the objective notion of evidence, this expert-relative view is incompatible with the historical thesis. Even on this view, which is relativized to what is known by the experts, e can be evidence for h regardless of whether any person knows that e is true. For example, given that current medical knowledge includes knowledge that spots of a

certain kind are always accompanied by measles, the presence of
these spots on John's body (e) is a reason to believe that John has the
measles (h). It is evidence for h because the fact that he has those
spots (e), given what is now noncontroversially known about such
spots (b), gives a reason for believing that h is true. This is so even if
there is no person, as yet, who knows that John has the spots. Hence,
this e is evidence even if no person knows that e is true. Moreover, e
can also be evidence on the expert-relative view if e *is* known to be
true (either by one person or by "the experts"). For example, prior to
Newton's formulation of his law of universal gravitation, it was
known that unsupported bodies fall. This information e was part of b
(it was known by the experts); still, this e was evidence for Newton's
universal theory. On the expert-relative theory of evidence, the time
at which e is known to be true by anyone or even by everyone is not
relevant to its status as evidence for h.

SCIENTIFIC EVIDENCE AND THE HISTORICAL THESIS

I have shown that while the historical thesis is compatible with a
personal theory of evidence, such as the subjective Bayesian view,
any purely personal theory is seriously flawed in not capturing the
nature of evidence as it is used in science. In science, evidence is an
impersonal concept. On impersonal concepts of evidence, such as the
expert-relative view and the objective views, some e can be evidence
for h whether or not any person knows or believes that e is true.
Hence the *time* at which e is known is not relevant to whether e is
evidence for h, as the historical thesis holds.

Some might argue that e cannot be evidence for any hypothesis if
no one knows that e is true, and might suggest prima facie rejection
of theories of evidence which entail this conclusion. This response,
however, conflates the distinction between *being* evidence and being
used as evidence. Suppose a doctor wants to consider the hypothesis
"this patient has AIDS." The doctor must decide whether or not there
is evidence for this hypothesis; that is, whether there is a reason to
believe that the patient has AIDS. The doctor then performs a blood
test which discovers the presence of the HIV virus. While it is still
not *certain* that the patient does have AIDS (the virus is present
before AIDS develops, and in fact not all infected people ever develop
AIDS), the doctor clearly does have some reason for believing that
the person has AIDS; and this reason exists even before the doctor
realizes that it does. Of course, it is only after the presence of the

virus is detected that the doctor *knows* there is evidence, or can *invoke* this fact as evidence, but these are different matters.

This is why in science, as in criminology, we speak of *discovering*— rather than inventing or creating—evidence. When a detective finds a suspect's fingerprints at the crime scene, we say that she has *discovered* evidence (provided, of course, that she has not planted the prints there herself. In this case, we say she has fabricated evidence, not that she has found evidence.) Similarly, scientists discover evidence that existed before it was found. Given the presence of HIV in a patient's blood there is a reason to believe the patient has AIDS, just as the presence of the suspect's fingerprints on the bloody chainsaw near the decapitated body is a reason to believe the suspect committed the murder, even before the virus or the prints are discovered. This is why detectives can speak of destroying evidence by handling objects before they are dusted for fingerprints; that is, destroying the evidence that exists but needs yet to be discovered. On the other hand, scientists can also discover an evidential relation between some information which was previously known, and a hypothesis. For example, doctors discovered that certain kinds of spots, which were already known to occur, were evidence that a patient has measles. If some information gives a reason to believe h, then it is evidence for h whether it is discovered to be true after h is invented or before it is—whether e is new evidence or old.[23]

I will close by paraphrasing Gertrude Stein: "evidence is evidence is evidence." The time at which some information is known relative to the forming of a theory is as irrelevant to its evidential value as is the time of day a rose is smelled to its status of being a rose.[24]

NOTES

1. Wesley Salmon (1966) similarly argues that, when a "daring" prediction is successful, ". . . it is not likely to come out right unless we have hit upon the correct hypothesis. Confirming instances are not likely to be forthcoming by sheer chance" (119).

2. John Worrall (1989) uses a version of this "no coincidence" argument in presenting his "heuristic" theory of evidence. A theory's empirical success, he claims, can be explained either by chance (an option Worrall rejects), by the theory's truth, or by the theory being "engineered" by the ingenuity of its creator. His heuristic approach is intended to preclude the third option, so that the only plausible option is that the theory is true. My arguments against views requiring either new or old evidence likewise work against Worrall's

theory, insofar as it considers relevant the time at which e was known by h's creator relative to h's invention. See also Gardner, 1982.

3. See Popper (1959), especially 40–42. In a related vein Giere has more recently argued that "if the known facts were used in constructing the model and were thus built into the resulting hypothesis. . . then the fit between these facts and the hypothesis provides no evidence that the hypothesis is true [since] these facts had no chance of refuting the hypothesis" (1984, 161).

4. Indeed, the more unlikely the prediction is, the more it increases the probability of the hypothesis when it is confirmed; that is, the smaller the figure in the denominator [p(e)], the larger p(h/e). Hence, successful predictions of surprising phenomena (what is contrary to expectation) count as the best evidence for a theory, on the positive-relevance view.

5. See Brush (1989), (1990), (1992), and (1993).

6. Brush argues for both a historical and a logical claim. The historical claim is that scientists do, in fact, prefer explanations to predictions as evidence. But he also presents a logical claim aimed against "Popper's thesis": that is, Brush argues that scientists *should* prefer explanations, because only explanations can be "reliable" evidence.

7. But note that there are other important reasons why the 1919 eclipse results may not have been "reliable" evidence. For one thing, the values obtained scatter around Einstein's predicted value, with probable errors large enough to include it; hence it is not obvious that the results *do* confirm the exact prediction. It is telling in this regard that it took four months of calculations before the scientific team making the observations concluded that the results *did* confirm the prediction. The eclipse occurred May 29; Einstein was not notified of the success until September 27; and the results were not announced to the public media until early November.

8. For discussions of "historical" views of confirmation, see Musgrave (1974) and McMullin (1979).

9. I am agreeing with the intuitive notion of evidence put forth by Achinstein (1983), although not with the details of his particular theory.

10. "Reasonableness to believe" is generally, and I think correctly, linked to the probability or the likelihood of the hypothesis. If e makes it reasonable to believe h, then h is very probable, given e. Proponents of the positive-relevance view discussed in the first section argue, by contrast, that e must increase the probability of h, and does not need to make h very probable.

11. *Science,* 260 (28 May, 1993), 1255–6, 1273–8.

12. To deny this role of evidence, and claim that scientists cannot use evidence as a means of deciding among theories, would be to claim that science is nonrational; that is, that scientists must always choose which theories to believe on the basis of social, political, and/or rhetorical reasons which have nothing to do with the attempt to discover how the world really works. While it may be true that scientists sometimes do, in fact, choose among theories in this nonrational way, it surely goes too far to claim that they do and should *only* choose which theories to believe on the basis of nonrational considerations.

13. See the following articles collected in Achinstein (1983): Carl G. Hempel, "Studies in the Logic of Confirmation," 10–43; Rudolf Carnap, "The Concept of Confirming Evidence," 79–94; Clark Glymour, "Relevant Evidence," 124–44; and Peter Achinstein, "Concepts of Evidence," 145–74.

14. Note that these are Achinstein's conditions for *potential* evidence.

15. For a full discussion of the subjective Bayesian position, see Howson and Urbach (1989).

16. See Glymour (1980), 85–93.

17. However, Howson and Urbach (1989) have argued that the predictivist position is not entailed by their view, because the conditional probability p(e/h) should be set for the counterfactual situation where one does not know, in advance, that e is true. Using this value of p(e/h) to determine p(h/e), the evidential value of e for h measures the extent to which e *would* increase a person's degree of belief. So on their construal of the subjective Bayesian position, the historical thesis is rejected (see 270–75).

18. The Bayesians claim that, in fact, people do come to have the same or similar degrees of belief in certain hypotheses; as more confirming instances of a universal law are observed, it becomes more likely that we will agree in expecting the next instance to also be a confirming one (there will be a "convergence of belief"). But this probabilistic claim has been defeated by John Earman (see his 1985). Moreover, our shared belief would extend only to the next instance, not to the belief in the truth of the hypothesis itself.

19. See *Principia,* Book III. For a more lengthy discussion of Newton and "the phenomena," see Achinstein (1991), 33–5.

20. To flesh out this view further, it would be necessary to explain how exactly the experts are to be distinguished from the nonexpert nonbelievers. That is, we need a way to determine when disagree-

ment signals ongoing controversy among scientific experts (as in the AIDS research community), and when it signals merely the disagreement of the nonexpert (as when the Church continued to deny the findings of Galileo, even after scientific controversy ended). This could be done to some extent by examining publications in respected scientific journals, membership in certain associations, etc.

21. This view is not to be confused with the standard way of relativizing evidential claims to "background information." When objective theories relativize claims to background information, as Carnap, Achinstein, and others do, this information b is taken to consist in any proposition or set of propositions at all—it is not limited to what is known by anyone, as I am suggesting here. When the subjective Bayesian view is relativized to background information, b consists in the knowledge and beliefs of a particular person; this also differs from relativization to a "total body of knowledge" of a community of experts.

22. The observed straight path of the landing stone is due not to the earth's immobility, but (as we now know) to inertia—to the tendency of a body to persist in a motion it has acquired. The stone continues to move eastward with the earth even when it is tossed up a few feet.

23. As the decapitated head example suggests, some e can be evidence for h even if it is neither explained nor entailed by h. That is, evidence does not *need* to consist in either an explanation or a prediction. From the hypothesis that John is dead, it does not follow as a predicted consequence that John's body is in two pieces. And the hypothesis that John is dead does not explain the fact that his head is no longer attached to his body (although it is certainly consistent with this fact). Yet the fact that his body is in this state is about as good a reason as we could have to believe that John is dead.

24. I am grateful to Peter Achinstein for many helpful suggestions. I would also like to thank Robert Rynasiewicz for comments on an earlier draft of this paper.

REFERENCES

Achinstein, P. (ed.) 1983. *The Concept of Evidence.* New York: Oxford University Press.

———. 1991. *Particles and Waves.* New York: Oxford University Press.

Brush, S. G. 1989. "Prediction and Theory Evaluation: The Case of Light Bending," *Science* 246: 1124–29.

————. 1990. "Prediction and Theory Evaluation: Alfven on Space Plasma Phenomena," *Eos* 71 (2): 19–33.

————. 1992. "How Cosmology Became a Science," *Scientific American* 267 (2): 62–70.

————. 1993. "Prediction and Theory Evaluation: Cosmic Microwaves and the Revival of the Big Bang," *Perspectives on Science* 1: 565–602.

Earman, J. 1985. "Concepts of Projectibility and Problems of Induction," *Nous* 19: 521–35.

Gardner, M. R. 1982. "Predicting Novel Facts," *British Journal for the Philosophy of Science* 33: 1–15.

Giere, R. N. 1984. *Understanding Scientific Reasoning,* 2nd edition. New York: Holt, Rinehart and Winston.

Glymour, C. 1980. *Theory and Evidence.* Princeton: Princeton University Press.

Gooding, D. et al. (eds.) 1989. *The Uses of Experiment.* Cambridge: Cambridge University Press.

Harman, G. 1965. "The Inference to the Best Explanation," *Philosophical Review* 74 (1): 88–95.

Howson, C. and P. Urbach. 1989. *Scientific Reasoning: The Bayesian Approach.* LaSalle, IL: Open Court Press.

McMullin, E. 1979. "The Ambiguity of Historicism," *Current Research in Philosophy of Science:* 55–83.

Musgrave, A. 1974. "Logical versus Historical Theories of Confirmation," *British Journal for the Philosophy of Science* 25: 1–23.

Newton, I. 1966. *Mathematical Principles of Natural Philosophy* (the *Principia*), 2 vols., trans A. Motte, rev. F. Cajori. Berkeley and Los Angeles: University of California Press.

Popper, K. R. 1959. *The Logic of Scientific Discovery.* London: Hutchinson and Co.

Salmon, W. 1966. *The Foundations of Scientific Inference.* Pittsburgh: University of Pittsburgh Press.

Whewell, W. 1857. *History of the Inductive Sciences,* 3rd edition, in three volumes. London.

————. 1860. *On The Philosophy of Discovery.* London.

Worrall, J. 1989. "Fresnel, Poisson and the White Spot: The Role of Successful Predictions in the Acceptance of Scientific Theories," in Gooding et al.

6

Art and Science: The Method of Ruskin's *Modern Painters*

Jonathan Smith

ABSTRACT

In historical studies of scientific methodology, we must take into account the cultural factors influencing methodological pronouncements. In nineteenth-century Britain, Baconian induction continued to enjoy considerable status as the *scientific method even as many cultural figures, including but not limited to scientists, redefined or rejected its attitude toward hypothesis. This paper examines a conflict between these two views in the work of John Ruskin (1819–1900), the Victorian art and social critic. I demonstrate the cultural power of induction by showing its importance as the methodological foundation of Ruskin's art criticism in his five-volume treatise on landscape art,* Modern Painters *(1843–60). Ruskin's later criticisms of science are thus shown to be primarily methodological, based on the inductive methods adopted in* Modern Painters, *and aimed particularly at John Tyndall's influential and (in Ruskin's view) insufficiently inductive "scientific use of the imagination."*

At the meeting of the British Association for the Advancement of Science in Liverpool on 16 September 1870, the physicist John Tyndall delivered an address entitled "The Scientific Use of the Imagination." In it, he celebrated the importance for the scientist of hypothesis, creative guesswork, and even intuition. The popular press responded angrily. What we admire in the scientific process, wrote the *Saturday Review,* is "the splendid series of inductions verified step after step by rigorous experiment and observation," the "firmly balanced and duly graduated tread of a mind trained in the discipline of logic, careful to plant every step on the assured ground of fact or experience," *not* "the leap of the imagination." (Tyndall 1870, 2–3).

In historical analyses of scientific methodology it is easy to assume that methodological pronouncements are of interest only to closed communities of scientists and philosophers. But as the reaction to Tyndall suggests, such an assumption cannot be made in nineteenth-century Britain. Indeed, many works on scientific method—the most influential including John Herschel's *Preliminary Discourse on the Study of Natural Philosophy* (1830), William Whewell's *Philosophy of the Inductive Sciences* (1840), J. S. Mill's *Logic* (1843), and the various translations and distillations of Auguste Comte's *Positive Philosophy*—were available for the generally educated reader, and numerous review articles on these works figured prominently in major periodicals. Public lectures were also much in vogue: at the Royal Institution in London, scientists like Humphry Davy, Michael Faraday, T. H. Huxley, and Tyndall all spoke on the methods as well as the content of contemporary science. Accounts of these lectures—often, as in the case of Tyndall's address, with editorial commentary—could be found not only in magazines like the *Saturday Review* but even in daily newspapers like the London *Times.*

The *Saturday Review*'s reaction to Tyndall also suggests, however, that the popular mind sensed a challenge to what it saw as the "approved" methodology of science: induction. When discussing this popular reaction and the cultural manifestations of its influence, whether on scientists or nonscientists, we need not necessarily ask if such a methodology actually existed in the past. The important historical fact, from a cultural standpoint, is simply that the *Saturday Review* and others said that it did exist. What Francis Bacon or Isaac Newton actually said about scientific method, in other words, is less important for a study of nineteenth-century culture than what nineteenth-century commentators *claimed* Bacon and Newton said. And rightly or wrongly, naively or not, for the *Saturday Review,* the method of science consisted of a series of inductions, not imaginative leaps. By the time of Tyndall's address, the English mind had absorbed for almost two centuries the lesson, admittedly simplistic, that what made science different from other intellectual pursuits was its method, that this method was induction, and that this induction, first articulated by Bacon, had made possible the spectacular successes of Newton. It was into this cultural environment that John Ruskin (1819–1900), the great Victorian critic of art and society, was born.

As a child and young man, Ruskin loved poetry and geology.[1] He published the first of many articles on geology at the age of fifteen in *The Magazine of Natural History*. As an undergraduate at Oxford, he won the Newdigate Prize for poetry and studied under the geologist William Buckland. Not surprisingly, even when his interests turned primarily to the criticism of art and society, he continued to stay abreast of contemporary science, applying it in his work, injecting himself into its controversies, and critiquing its place in the culture. Today, Ruskin's views on science are often dismissed as the ravings of an incipient madman or minimized as almost willfully out of touch. Yet his objections to specific scientific theories were frequently made on *methodological* grounds. And Ruskin's methodology, consistently employed throughout forty years of study in both art and science, was not the method of Tyndall's scientific imagination but of the *Saturday Review*'s rigorous induction. To examine Ruskin's method is thus to develop an appreciation of the importance of what we might call the cultural factors influencing the selection and articulation of any scientific methodology in nineteenth-century Britain.

Ruskin's *Modern Painters* appeared in five volumes between 1843 and 1860. It began as, and remained throughout, a defense of the British landscape artist J. M. W. Turner, whose works were savagely criticized in the 1830s and 1840s as unfaithful representations of nature. Turner's landscapes, unlike those of such Old Masters as Claude Lorrain and Gaspard and Nicolas Poussin, were said to be subjective and impressionistic, full of fabulous colors, impossible lighting effects, and fanciful natural forms. Ruskin, on the contrary, argued that art critics had derived their conception of nature from the landscapes of the Old Masters rather than from nature itself. But *Modern Painters* is a far more ambitious project than a defense of Turner, for Ruskin recognized quickly that such a defense required a statement of his own aesthetic principles in the context of a broader survey of Western landscape art. There is, inevitably, a certain amount of change in Ruskin's specific views over the course of seventeen years, a fact Ruskin himself acknowledges in the Preface to the final volume. But in that same Preface he also claims that the methodology governing his criticism is consistent: "In the main aim and principle of the book, there is no variation, from its first syllable to its last."[2]

What was this methodology? In *Modern Painters I*, Ruskin argues that the landscape painter "must always have two great and distinct ends":

the first, to induce in the spectator's mind the faithful conception of any natural objects whatsoever; the second, to guide the spectator's mind to those objects most worthy of its contemplation, and to inform him of the thoughts and feelings with which these were regarded by the artist himself. (3:133)

For Ruskin, great landscape art begins with "the faithful conception of natural objects." The artist's emotional and intellectual response to the landscape is of no value without this careful representation of visible nature:

[A]lthough it is possible to reach what I have stated to be the first end of art, the representation of facts, without reaching the second, the representation of thoughts, yet it is altogether impossible to reach the second without having previously reached the first. . . . And this is the reason why, though I consider the second as the real and only important end of all art, I call the representation of facts the first end; because it is necessary to the other and must be attained before it. It is the foundation of all art. (3:136)

Notice that Ruskin does not limit art to "the representation of facts." This would be mere copying or imitation, which for Ruskin produces pleasures "the most contemptible which can be received from art" (3:101). Great art, on the contrary, demands "the representation of thoughts," the presence of an active artistic imagination selecting and rearranging its subject matter. But such selection and manipulation must be founded on the faithful representation of the facts of nature.

Ruskin's strategy for responding to Turner's critics is thus to show that Turner is not less but more faithful to nature, that Turner looks at nature with greater care than any landscapist living or dead. By contrast, the works of the Old Masters only seem to be faithful to nature because we ourselves have not observed nature closely. Over and over again, Ruskin stresses that seeing must actively involve both the senses and the mind: "unless the minds of men are particularly directed to the impression of sight, objects pass perpetually before the eyes without conveying any impression to the brain at all; and so pass actually unseen, not merely unnoticed, but in the full clear sense of the word unseen" (3:142). This passivity of vision on the part of painters leads to unfaithful representations of nature that, in the hands of equally passive critics, become the standard for truth. Of Italian skies, for example, Ruskin writes:

How many people are misled, by what has been said and sung of the serenity of Italian skies, to suppose they must be more *blue* than the skies of the north, and that they see them so; whereas the sky of Italy is far more dull and grey in colour than the skies of the north, and is distinguished only by its intense repose of light. . . . And what is more strange still, when people see in a painting what they suppose to have been the source of their impressions, they will affirm it to be truthful, though they feel no such impression resulting from it. Thus . . . they will affirm a blue sky in a painting to be truthful, and reject the most faithful rendering of all the real attributes of Italy cold or dull. (3:144)

For Ruskin, a "faithful rendering" of natural objects involves the representation of what the artist actually sees, which in turn guarantees fidelity to nature. So insistent is Ruskin on this point that he warns painters to avoid the common error of "suppos[ing] that they see what they know" to exist, and then "of painting what exists, rather than what they can see" (3:144–45). The representation of natural objects must be founded on close observation uncluttered by the reason or the imagination, by what one knows or what one hopes to be there. Certain pine trees, for example, are dark green, but when seen from a distance their color may be tinted with purple. To be faithful to nature, the landscapist must paint them dark green if he is close enough to see them that way, but he must paint them with a purple tint if he is painting from a distance. Just because he knows the trees are dark green when seen up close does not mean he should paint them that way. As Patricia Ball (1971, 65) puts it:

[S]eeing is for [Ruskin] the prime qualification of the great artist. Emotion and further enlightenment depend upon the completeness of the encounter with the object. If it is only vaguely seen, manipulated for the artist's convenience, or reduced to his preconceived assumptions about it, there will be no revelation, no noble emotion. Turner is to be acknowledged as a hyper-sensitive eye before he can be hailed as nature's interpreter. That role arises out of his subjection to the snowstorm, the cloud or the sea; he does not begin from his emotions about these things. The facts come first.

In keeping with both his methodology and his argument, Ruskin devotes the bulk of *Modern Painters I* to a demonstration of Turner's faithful renderings of nature, first in the more general truths of tone, color, and space, and then in the specific truths common to landscapists of skies, clouds, rocks, earth, mountains, water, and vegetation.

As an observer of nature, Ruskin is extraordinarily precise and detailed, offering pages of description of natural phenomena. In his discussion of one of the "common and general optical laws which are to be taken into consideration in the painting of water" (3:508), for example, Ruskin describes the various reflections that can be seen in different parts of the rippled surface of a body of water:

If water be rippled, the side of every ripple next to us reflects a piece of the sky, and the side of every ripple farthest from us reflects a piece of the opposite shore, or of whatever objects may be beyond the ripple. But as we soon lose sight of the farther sides of the ripples on the retiring surface, the whole rippled space will then be reflective of the sky only. Thus, where calm distant water receives reflections of high shores, every extent of rippled surface appears as a bright line interrupting that reflection with the colour of the sky.

When a ripple or swell is seen at such an angle as to afford a view of its farther side, it carries the reflection of objects farther down than calm water would. Therefore all motion in water elongates reflections, and throws them into confused vertical lines. The real amount of this elongation is not distinctly visible, except in the case of very bright objects, . . . whose reflections are hardly ever seen as circles or points, which of course they are on perfectly calm water, but as long streams of tremulous light. (3:506)

This "optical law" is the result of careful observation, and such descriptions actually serve as the basis for Ruskin's analysis of the depiction of these phenomena in individual paintings. In his critique of Claude Lorraine's "Sea-piece, with a Villa," for example, Ruskin complains that Claude's representation of the sun's reflection is simply wrong:

The sun is setting at the side of the picture, it casts a long stream of light upon the water. This stream of light is oblique, and comes from the horizon, where it is under the sun, to a point near the centre of the picture. If this had been done as a license, it would be an instance of most absurd and unjustifiable license, as the fault is detected by the eye in a moment, and there is no occasion or excuse for it. But I imagine it to be an instance rather of the harm of imperfect science. Taking his impression instinctively from nature, Claude usually did what is right and put his reflection vertically under the sun; probably, however, he had read in some treatise on optics that every point in this reflection was in a vertical plane between the sun and the spectator; or he might have noticed, walking on the shore, that the reflection came straight from the sun to his feet, and intending to indicate the position of the spectator, drew in his next picture the reflection sloping to this supposed point . . . . (3:511)

Ruskin is willing to give Claude the benefit of the doubt: the error is probably the result not of "license" but of "imperfect science," and imperfect science is itself the result of careless observation or of theorizing without observation. Indeed, Ruskin goes on to remark that Claude's error is "plausible enough to have been lately revived and systematized" (3:511n) in books on geometric optics and perspective, and his brief refutation of this error is instructive.

Admitting that he lacks the space to enter into a lengthy disputation, Ruskin notes that

> reasoning is fortunately unnecessary, the appeal to experiment being easy. Every picture is the representation, as before stated, of a vertical plate of glass, with what might be seen through it drawn on its surface. Let a vertical plate of glass be taken, and wherever it be placed, whether the sun be at its side or at its centre, the reflection will always be found in a vertical line under the sun, parallel with the side of the glass. The pane of any window looking to sea is all the apparatus necessary for this experiment; and yet it is not long since this very principle was disputed with me by a man of much taste and information, who supposed Turner to be wrong in drawing the reflection straight down at the side of his picture, as in his Lancaster Sands . . . . (3:511n)

As is typical of him, Ruskin is not hostile to science but to what he sees as bad science. Claude has been led astray by those who have reasoned without looking, so Ruskin reverses the process, rejecting reasoning in favor of a simple visual experiment. And this exercise in careful observation reveals that Turner's work, though seemingly at odds with both nature and optics, is in fact in closer accord with both than Claude's.

This methodology is central not simply to *Modern Painters* but virtually to all of Ruskin's varied publications. In particular, it was the basis for his own works of science and for his criticism of many contemporary scientific theories. As Ruskin himself remarked in a paper for the Mineralogical Society nearly forty years after the appearance of *Modern Painters I,* "precisely the same faculties of eye and mind are concerned in the analysis of natural and of pictorial forms" (26:386).

Ruskin's modern critics, even his most sympathetic ones, while acknowledging this similarity in approach to "natural and pictorial forms," have tended to see Ruskin as badly out of step with the science of his own day. John Rosenberg, for example, refers to Ruskin's geological essays of the 1870s and early 1880s as "capricious" and "deliberately unscientific" (1961, 180). Rosenberg argues that Ruskin turns to such studies of nature, undertaken during a period of depression and breakdown, to ward off madness. Ruskin's geology

> was a combination of lifelong study of the Alps and pleasant putterings in his kitchen with toast-crumb morraines and glaciers of blancmange. It is a measure of his increasing intellectual isolation that his master in geology was Saussure, whose *Voyages dans les Alpes* had first appeared in 1779. With defensive arrogance, he attacked the contribution of contemporaries, never quite realizing that his own was not science but play. (1961, 180)

Yet Ruskin's "isolation" has much less to do with the *content* of contemporary geology than it does with its *methodology*. In his extensive discussion of Alpine geology in *Modern Painters IV* (1856), Ruskin notes that "the natural tendency of accurate science is to make the possessor of it look for, and eminently see, the things connected with his special pieces of knowledge; and as all accurate science must be sternly limited, his sight of nature gets limited accordingly" (6:475).[3] In contrast, his own approach to natural objects is to remain "as much as possible in the state of an uninformed spectator of the outside of things, receiving simply what impressions the external phenomena first induce" (6:476). "I closed all geological books," writes Ruskin, "and set myself . . . to see the Alps in a simple, thoughtless, and untheorizing manner; but to *see* them, if it might be, thoroughly" (6:476). Saussure is the only geologist whose help he has received because

> all other geological writers whose works I had examined were engaged in the maintenance of some theory or other, and always gathering materials to support it. But I found Saussure had gone to the Alps, as I desired to go myself, only to look at them, and describe them as they were, loving them heartily . . . more than himself, or than science, or than any theories of science . . . . (6:476)

Saussure's verbal representations of the Alps, like Turner's pictorial representations of them, are based on careful and theoretically neutral observation of the visible. For Ruskin, reason and imagination follow this process of observation; in most nineteenth-century geology, on the contrary, they have preceded it. Thus in *Modern Painters* Ruskin utilizes his own drawings and observations of Mont Blanc to challenge the "chief theory with geologists" for its formation (6:253), while one of Turner's Alpine paintings serves as the basis for

a description of the laws of rock cleavage (6:269–75). In *Deucalion,*
his collection of geological essays, Ruskin's stance is similar. "The
theories respecting the elevation of the Alps," he writes, remain
"uncertain and unsatisfactory," and therefore our own work

> must waste no time on them; we must begin where all theory ceases; and
> where observation becomes possible,—that is to say, with the forms which
> the Alps have actually retained while men have dwelt among them, and
> on which we can trace the progress, or the power, of existing conditions of
> minor change . . . . I do not care . . . how crest or aiguille was lifted, or
> where its materials came from, or how much bigger it was once. (26:112–
> 13)

Ruskin was a relentless critic of contemporary science whenever
he saw it relying on theory over observation, trying to explain
unobserved or unobservable phenomena when observable phenomena remained shrouded in obscurity, or using terminology that
created the illusion of explanation.[4] And Ruskin's chief scientific
villain was John Tyndall and his "scientific use of the imagination."
     Like Ruskin, Tyndall was a great lover of the Alps, traveling
through them frequently to study the same natural phenomena that
fascinated Ruskin. He reported the results of his research, particularly of his work on glaciers, in lectures at the Royal Institution, and
he wrote several books on the subject, including *The Glaciers of the
Alps* (1860) and *The Forms of Water in Clouds and Rivers, Ice and
Glaciers* (1872). In these books and lectures, however, Tyndall
engaged in an acrimonious dispute about glacier motion with J. D.
Forbes, whose own theory was first published in *Travels Through the
Alps* in 1843. Ruskin had met Forbes in 1844 and considered his book
the definitive work on Alpine glaciers, for Forbes, like Saussure, had
taken the trouble to observe, and had only built his theory on the
foundation of what he had seen. Indeed, Ruskin quotes Forbes
copiously in his discussion of the central Alps in *Modern Painters IV,*
even according Forbes pride of place over Saussure. In the ensuing
controversy, which extended well beyond Forbes's death in 1868,
Ruskin was active in defense of Forbes.[5] He lectured on glaciers
himself at the Royal Institution in 1863, championing Forbes over
Tyndall, and repeatedly introduced the subject into his various
works, including *Deucalion* and the new editions of *Modern Painters.*
     While the philosophical differences between Ruskin and Tyndall
were wide, the bulk of Ruskin's criticisms focuses on Tyndall's
method, his alacrity in spinning out theories grounded not in the use

of his eyes but of his imagination. Comparing sketches by both Forbes and Tyndall of the Mer de Glace glacier in *Deucalion,* Ruskin sarcastically offers the most damning criticism he can make: that "Professor Tyndall is unable to draw anything as seen from anywhere, . . . such inability serving farther to establish the sense of his proud position as a man of science, above us poor artists, who labour under the disadvantage of being able with some accuracy to see, and with some fidelity to represent, what we wish to talk about" (26:161).

Tyndall's inaccurate vision and unfaithful representations, Ruskin later argues, are both a product of, and a contributor to, bad theories and bad explanations. In a discussion of the formation of ice on the surface of calm water, Ruskin notes that the hexagonal structure of the ice crystals is "sufficiently explained, in Professor Tyndall's imagination, by the poetical conception of 'six poles' for every hexagon of ice" (26:358). In *The Forms of Water,* Tyndall had claimed this as a triumphant example of the scientific use of the imagination:

> Our first notions and conceptions of poles are obtained from the sight of our eyes in looking at the effects of magnetism, and we then transfer these notions and conceptions to particles which no eye has ever seen. The power by which we thus picture to ourselves effects beyond the range of the senses is what philosophers call the Imagination, and in the effort of the mind to seize upon the unseen architecture of crystals, we have an example of the 'scientific use' of this faculty. (1872, sec. 95)

Tyndall contends that by *seeing* the effects of magnetism he is then able to explain the form and cause of the *unseen* ice crystals; by analogy, each hexagonal crystal thus consists of six "poles." Ruskin, of course, sneers at such an "explanation":

> Perhaps!—if one knew first what a pole was, itself—and how many, attractive, or repulsive, to the east and to the west, as well as to the north and the south—one might institute in imaginative science—at one's pleasure;—thus also allowing a rose five poles for its five petals, and a wallflower four for its four, and a lily three, and a hawkweed thirteen. (26:359)

Tyndall is no Turner. He can't draw because, unlike Turner, he leaps too rapidly from observation to imagination. And Tyndall's language gives him away: rather than providing what Ruskin calls in *Modern Painters* "a downright statement of facts" (3:138), it is vague, theoretical, and metaphorical.

In *Ruskin and the Art of the Beholder,* Elizabeth Helsinger argues that Ruskin's theories of perception sought to correct, on the one hand, the scientist's failure to take into account his response to nature, and on the other hand, the poet's tendency to see nature subjectively, exclusively through his own emotional state—what Ruskin christened the "pathetic fallacy." According to Helsinger, the Ruskinian "beholder" is a "wandering natural scientist" who, "like [the] visionary artist, does not merely observe but perceives: his own responses, shaping his observations, are an admitted part of his subject" (1982, 64). Applying Helsinger's argument to Tyndall, we must ask why Ruskin was so critical of this particular "wandering natural scientist."

In *Modern Painters III,* Ruskin describes the poet who succumbs to the pathetic fallacy as one "who perceives wrongly, because he feels, and to whom the primrose is anything else than a primrose: a star, or a sun, or a fairy's shield, or a forsaken maiden" (5:209). The language, including the floral example, recalls the passage above in which Ruskin sneers at applying the term "poles" to a crystal of ice or the petals of a flower. More importantly, it suggests that Tyndall's scientific use of the imagination results in the scientific version of the pathetic fallacy. Tyndall's failure, in Ruskin's eyes, is not that he lacks an emotional response to nature but that his emotional response is too strong. His language expresses admiration for the "architecture" of the ice crystal and reverence for the science that can explain such natural beauties by going "beyond the range of the senses." This emotional response leads Tyndall to employ his imagination to construct a scientific explanation for the crystal. But just as the primrose is anything else than a primrose for the poet who feels, the ice crystal is anything else than an ice crystal for Tyndall— it is a magnet with six poles. In Ruskin's view *both* poet and scientist are susceptible to the pathetic fallacy, and in both cases the result is an inaccurate perception of nature.

Ruskin's hostility to the vague language of scientific explanations and descriptions also surfaced in his encounters with the wave theory of light and the kinetic theory of matter. In the early 1880s, Ruskin again became interested in cloud phenomena, reprinting the cloud sections of *Modern Painters IV* and *V* as a separate volume and delivering two lectures at the Royal Institution on "The Storm-Cloud of the Nineteenth Century." In the first "Storm-Cloud" lecture, Ruskin assesses the current explanations of such phenomena, opening his discussion with "a little word" of "extremely useful advice about scientific people in general" (34:17). Their business, he says, is to tell us that certain things are so, that they do happen, and then to tell us what to do if we want these things to happen or not. But, "if . . . they ever try to *explain* anything to you, you may be confident of one of two things,—either that they know nothing (to speak of) about it, or that they have only seen one side of it—and not only haven't seen, but usually have no mind to see, the other" (34:17). Ruskin's rhetoric is inflated—he really does want to be able to explain the cause of these phenomena—but he complains that he is constantly "stopped dead" by the scientists' "confusion of ideas" in "using the words undulation and vibration as synonyms":

> "When," says Professor Tyndall, "you are told that the atoms of the sun vibrate at different rates, and produce waves of different sizes,—your experience of water-waves will enable you to form a tolerably clear notion of what is meant."
> "Tolerably clear"!—your toleration must be considerable, then. Do you suppose a water-wave is like a harp-string? (34:25–26)

Ruskin's criticism is off the mark: Tyndall is not using vibration and undulation as synonyms but claiming that the vibrations of the sun's atoms produce waves in the surrounding ether. Ruskin's ultimate point, however, is valid. Water waves are physical objects, but "the undulating theory of light is proposed to you concerning a medium [the ether] which you can neither see nor touch" (34:26). Ruskin, interestingly, is consistent with his own methodology, pointing out that his criticism does not imply a rejection of the ether because there is no positive evidence to show that the ether does *not* exist. Rather, he is an agnostic about the ether, the wave theory of light, and kinetic theories of matter: "I neither accept, nor assail, the conclusions respecting the oscillatory states of light, heat, and sound, which have resulted from the postulate of an elastic, though impalpable and imponderable ether, possessing the elasticity of air" (34:26–27). What he objects to is the way Tyndall's scientific imagination, operating through an analogy between water waves and light waves, treats the analogy as a description and an explanation of what light is, and hence as sufficient grounds for asserting the existence of an ether that can't be seen or felt or weighed.

Going to the heart of the wave theory, Ruskin also raises the question of how the sun's atoms, as "dead matter," are to start vibrating in the first place: "*You* may fall a-shivering on your own account, if you like, but you can't get a billiard-ball to fall a-shivering on *its* own account" (34:26). Ruskin was wrong here as well, but he retracted the statement in his second lecture, quoting a passage from a paper by G. G. Stokes on "epipolized" (fluorescent) light:

> "Nothing seems more natural than to suppose that the incident vibrations of the luminiferous ether produce vibratory movements among the ultimate molecules of sensitive substances, and that the molecules in return, *swinging on their own account,* produce vibrations in the luminous ether, and thus cause the sensation of light." (34:59)

Stokes, the Lucasian Professor of Mathematics at Cambridge, commanded great respect from Ruskin. The holder of Newton's chair, Stokes was not the materialist that Tyndall was (Wilson 1989). And the paper from which Ruskin quotes, "On the Change of Refrangibility of Light," consists primarily of the sort of careful and detailed experimentation and observation that Ruskin loved. But Stokes was an early proponent of the wave theory, and all his work on light was based on it. Ruskin is thus careful, in citing Stokes on the self-induced quality of the molecular vibrations, to begin his quotation *after* Stokes's endorsement of the wave theory and to ignore entirely Stokes's claim that the wave theory is to light what the theory of gravitation is to the heavenly bodies. While willing to grant that vibration can produce undulation (and hence that scientists are not using the terms synonymously), Ruskin does not want to include himself among those whom Stokes terms "believers in the undulatory theory of light" (Stokes 1901, 270, 388).

From our vantage point, Ruskin's ambivalence about Stokes's explanation of fluorescence and his agnosticism about the wave theory appear reasonable despite the hyperbole of his rhetoric. Although Stokes argues that "nothing seems more natural" than his explanation, fifteen years later he withdrew it for a different one he considered "far more probable" (3:388 and n). And the existence of the ether was of course ultimately disproven by the Michelson-Morley experiment in 1887, just three years after the "Storm Cloud" lectures. As in the case of glacier theory, then, Ruskin was quite up-to-date with the content of contemporary scientific theories of light and matter, and his objections to them were strongly rooted in methodological concerns. Indeed, from *Modern Painters I* in 1843 to the geological writings of the 1860s and 1870s and on to the "Storm-Cloud" lectures of 1884, Ruskin's approach to the representation of natural phenomena remained consistent. Ruskin could have easily characterized his methodology in the same words the *Saturday Review* employed in its own rejection of Tyndall, that the proper method of science is "the splendid series of inductions verified step after step by rigorous experiment and observation," the "firmly balanced and duly graduated tread of a mind trained in the discipline of logic, careful to plant every step on the assured ground of fact or experience."

But how are we to account for the rather odd fact that Ruskin, the "literary" and "artistic" figure, employs a more restrictive methodology than scientists like Stokes and Tyndall? As I suggested at the outset, discussions of scientific methodology were often carried on in public in nineteenth-century Britain—they were not merely the province of philosophically minded scientists and scientifically minded philosophers. Ironically, by 1850, science found itself in a methodological trap; on the one hand, its success in publicizing its inductive triumphs had made possible pronouncements like those of the *Saturday Review;* on the other hand, it wanted to prevent that view from becoming a naive and restrictive caricature that denied the importance of imagination and intuition, hunches and guesswork, to the working scientist. In 1808, Humphry Davy declared at the Royal Institution that the "legitimate practice" of science was "that sanctioned by the precepts of Bacon and the examples of Newton" (1839–40, 8:276). Twelve years later, just one year after the birth of Ruskin, Davy urged his colleagues in his inaugural address as President of the Royal Society to "be guided by the spirit of philosophy, awakened by our great masters, Bacon and Newton, that sober and cautious method of inductive reasoning, which is the germ of truth and of permanency in all the sciences" (1839–40, 7:14). Such public pronouncements were powerful. They help explain why the works on scientific method in the 1830s and 1840s, including Herschel's, Whewell's, and Mill's, were carefully presented as updating rather than altering the Baconian-Newtonian tradition. They help explain why Tyndall's views on "the scientific use of the imagination," often expressed in the same lecture theater of the Royal Institution, could meet with such popular criticism.[6] And they help explain why Ruskin, addressing a popular audience in the very same room, could make similar criticisms of Tyndall and not be thought a crank or an anti-scientific zealot.

As for why Ruskin was not in the methodological vanguard, a comparison with Stokes's life is instructive. The two men were born in the same year, but while Ruskin attended Oxford, the seat of classical learning, Stokes attended Cambridge, the home of Whewell and many of the wave theorists. When Ruskin was attending the geological lectures of Buckland, Stokes was studying Continental mathematical techniques and physical theories. Although Ruskin later met Whewell and developed a friendship with him—Whewell shared Ruskin's love of Gothic architecture and had even published a book on its development—Ruskin was consistent here as well, praising Whewell for his close observation of architectural details but never endorsing Whewell's general theories about Gothic, which clearly came before rather than after the close observation of details.[7]

Ruskin was, in other words, a product of his culture as well as a critic of it. And this culture included a strong strain of obeisance to the successes of the inductive methodologies of science. For Ruskin, the discovery of general laws, whether of "natural or pictorial forms," had to begin with careful, precise observation. As he once jokingly advised the physicist Oliver Lodge, in a conscious echo of Newton's famous "hypotheses non fingo," "waste no time in hypotheses; I never made but one in my life, and that was wrong. I only want to know what *is*" (37:526). He may have been naive, but Ruskin believed that it was possible to observe objectively, to determine what he called in *Modern Painters* a "downright statement of facts" (3:138) on which the imagination and reason could then, safely, build.

Indeed, for Ruskin, this is the point at which science and landscape art diverge. Science studies the physical causes of natural phenomena, while art examines their aesthetic and moral implications: "there is a science of the aspects of things, as well as of their nature; and it is as much a fact to be noted in their constitution, that they produce such and such an effect upon the eye or heart . . . as that they are made up of certain atoms or vibrations of matter" (5:387). In *Modern Painters III,* Ruskin declares that "the master of this science of *Aspects*" is Turner, who "must eventually be named always with Bacon, the master of the science of *Essence*" (5:387). Turner, says Ruskin, "must take place in the history of nations corresponding in art accurately to that of Bacon in philosophy;—Bacon having first opened the study of the laws of material nature, . . . and Turner having first opened the study of the aspect of material nature" (5:353).

Turner initiates a revolution in landscape art, a revolution based, like Bacon's, in the study of "material nature." In *Modern Painters* and throughout his long career, Ruskin remained committed to the inductive methodology that made both the Baconian and the Turnerian revolutions possible. This commitment simultaneously contributed to Ruskin's status as a revolutionary critic of art and architecture, put him in step with his culture's conception of scientific method, and led to disputes with scientists like Tyndall who were trying to revise that cultural conception. The case of Ruskin confirms the power of a strong inductive tradition within British culture, not merely within British science, and it suggests that the study of scientific methodology in the nineteenth century must always be sensitive to the power and influence of such cultural contexts and traditions. In our own historical studies of scientific method, we, too, must construct a science of aspects as well as a science of essence.

Notes

1. For biographical information on Ruskin, see Hunt (1982).

2. Ruskin 1903–12, 7:9. Subsequent references to Ruskin's *Works* will be given only by volume and page number.

3. Another of Ruskin's modern admirers, George Landow, says in his important study of Ruskin's aesthetics that this "long disquisition" on geology induces the "bored and bewildered" reader to "glance at the title page to reassure himself that he is still pursuing a work about Turner" (1971, 34). Landow, however, limits his focus to Ruskin's theories of beauty, a strategy that leads him to undervalue the importance of such disquisitions on geology to the proper appreciation of Turner.

4. Ruskin took great pleasure in poking fun at scientific terminology. In *Deucalion,* he complains that the "[g]reat part of the supposed scientific knowledge of the day is simply bad English, and vanishes the moment you translate it" (26:260). As an example, Ruskin notes the absurdity of "what you, in compliment to Greece, call a 'Dinotherium,' Greece, in compliment to you, must call a 'Nastybeastium,'— and you know that interchange of compliments can't last long" (26:261).

5. For a full account of the controversy, and Ruskin's role in it, see 26:xxxiii–xli.

6. It is worth keeping in mind that even in his *Autobiography,* Darwin claimed that in his research for the *Origin of Species* he had "worked on true Baconian principles, and without any theory collected facts on a wholesale scale" (1989, 144). Privately, however, Darwin consistently denied the practicality of collecting facts without at least a provisional theory for a guide.

7. Whewell's *Architectural Notes on German Churches* first appeared in 1830 but was enlarged and republished in 1835 and 1842

(Todhunter 1876, 1:42–44). Whewell's theories about architectural history, like his views of scientific methodology, were strongly influenced by German idealist philosophy. Ruskin, on the contrary, opens the chapter of *Modern Painters III* on the pathetic fallacy with a sharp attack on German philosophy. In letters to his niece on this volume of *Modern Painters,* Whewell responded to Ruskin's attack, referring somewhat caustically to "you Ruskinians" and complaining about Ruskin's "idolatry of Turner" (Douglas 1882, 464, 466, 476). Whewell also wrote a mixed review for *Fraser's Magazine* of Ruskin's *The Seven Lamps of Architecture* (1849), the work in which Ruskin puts some of Whewell's observations into the service of a rather different argument about Gothic architecture (Todhunter 1876, 1:173).

References

Ball, Patricia. 1971. *The Science of Aspects: The Changing Role of Fact in the Work of Coleridge, Ruskin, and Hopkins.* London: Athlone.

Darwin, Charles. 1989. *The Autobiography of Charles Darwin.* Vol. 29 of *The Works of Charles Darwin,* Paul H. Barrett and R. B. Freeman (eds.). London: Pickering.

Davy, Humphry. 1839–40. *The Collected Works of Sir Humphry Davy,* John Davy (ed.). 9 vols. London: Smith, Elder.

Douglas, Mrs. Stair (Janet Mary). 1882. *The Life and Selections from the Correspondence of William Whewell,* 2d ed. London: Kegan Paul.

Helsinger, Elizabeth K. 1982. *Ruskin and the Art of the Beholder.* Cambridge: Harvard University Press.

Hunt, John Dixon. 1982. *The Wider Sea: A Life of John Ruskin.* London: Dent.

Landow, George P. 1971. *The Critical and Aesthetic Theories of John Ruskin.* Princeton: Princeton University Press.

Rosenberg, John D. 1961. *The Darkening Glass: A Portrait of Ruskin's Genius.* New York and London: Columbia University Press.

Ruskin, John. 1903–12. *The Library Edition of the Works of John Ruskin,* E. T. Cook and Alexander Wedderburn (eds.). 39 vols. London: George Allen.

Stokes, George Gabriel. 1901. "On the Change of Refrangibility of Light." In vol. 3 of *Mathematical and Physical Papers.* 5 vols. Cambridge: Cambridge University Press.

Todhunter, Isaac. 1876. *William Whewell: An Account of His Writings*. 2 vols. London: Macmillan.

Tyndall, John. 1870. *Essays on the Use and Limit of the Imagination in Science*, 2d ed. London: Longmans.

———. 1872. *The Forms of Water in Clouds and Rivers, Ice and Glaciers*. New York: Appleton.

Wilson, David B. 1989. "A Physicist's Alternative to Materialism: The Religious Thought of George Gabriel Stokes." In *Energy and Entropy: Science and Culture in Victorian Britain*, Patrick Brantlinger (ed.), 177–204. Bloomington, IN: Indiana University Press.

7

NARRATIVE JUSTIFICATION IN PHILOSOPHY OF SCIENCE: A ROLE FOR HISTORY

D. Lynn Holt

History, if viewed as a repository for more than anecdote or chronology, could produce a decisive transformation in the image of science by which we are now possessed.

—Thomas Kuhn, *The Structure of Scientific Revolutions*

What would it be like to approach scientific inquiry from a Gassendist perspective, one which in addition to logic and experiment would provide historical arguments?

—Lynn Joy, *Gassendi the Atomist*

ABSTRACT

Many will grant that in practice there is no such thing as the history of science, that histories differ in emphasis and perspective. But few will concede that historical narrative can differ in principle to such a degree that one calls Aristotelian science a success and another calls it unscientific. Such radical difference reflects a difference of historiography. I outline two opposed ways of writing the history of science: one relies on standards of truth and method which are assumed to be timeless, the other employs standards which are indigenous to the period under study. I go on to show both ways mistaken, and suggest a third which overcomes the limitations of both while incorporating their best insights. Along the way, I seek to establish that narrative history is ultimately necessary for justifying scientific claims, thus making it no accident that logic and observation jointly fail to determine truth in science.

Suppose that the title "The Father of Modern Philosophy" were given to Pierre Gassendi instead of René Descartes. This would mean that the sixteenth- and seventeenth-century debate between antihistorical methodologists like Descartes and Bacon and late humanist historiographers like Gassendi had been publicly won by the historiographers. Imagine, as a consequence, that the official modern (and subsequent contemporary) view turned out to be not that some timeless combination of logic and observation justifies scientific practices and scientific theories, but rather that they are justified largely by historical narratives of victories, partial defeats and successive modifications in contests with their rivals. How would contemporary orthodox philosophy of science differ?

It would be premature to answer this question until both an understanding of narrative justification and an appreciation of the different forms which that justification may take has been reached. Accordingly, I will first identify two main forms which narrative justification might take, isolating the standards of evaluation appropriate to each, and giving examples of each. In these sections I hope to dispel the notion that history of science is one thing and philosophy of science quite another, and to show that history has an essential role to play in the justification of scientific practice and theory.

But these two dominant forms of historiography are not without their problems. Hence I will sketch an alternative to the main forms of narrative justification which I survey. Having done so, I will at last turn to the question of how philosophy of science would have been different, drawing on my arguments for why it should be different.

A final note. I take the lessons of historiography seriously, and so this essay is, at least in part, historically organized. To do otherwise would be to commit the same mistakes that I accuse others of. So the analytical devices which the reader expects—definitions, distinctions and so on—will be found not at the beginning, but in the course of my narrative. As with any good story, plot devices should emerge naturally, and then only when necessary.

HOW TO WRITE THE HISTORY OF SCIENCE I: AHISTORICISM AND THE TYRANNY OF THE PRESENT

René Descartes and Francis Bacon are justly regarded as founding figures in the development of the modern mind. For even though they differ on, among other things, the evidentiary role assigned to observation in the development of knowledge, they nevertheless agree on a key assumption which was perpetuated by eighteenth- and nineteenth-century British (and French) positivist empiricism and continues to dominate much contemporary Anglo-American philosophy of science under the rubric Logical Empiricism: the justification of scientific knowledge is ahistorical. That is, history plays no role in the justification of scientific knowledge; knowledge (whether an individual claim made by a scientist, or the entirety of the scientific project) is justified by some combination of logic and observation, broadly construed, and these will encompass timeless truths and/or methods. For Descartes, such knowledge is the product of logical derivation from the timeless truths of reason such as "Every effect has a cause"; for Bacon, such knowledge is the product of induction on timeless facts of observation such as "The heat in this piece of wood is accompanied by motion." But the secure foundation which these two different bases for knowledge are meant to afford derives in key part from the twofold assumption that these knowledge bases do not change over time, and that their justification is available to any person regardless of that person's historical standpoint. And if we identify scientific rationality with the justification of scientific knowledge, then the modern mind conceives of scientific rationality as ahistorical.

It is somewhat curious, then, if not downright paradoxical, that history of science could play any role in an ahistorical conception of science. But what I will call ahistoricism turns out to be a matter of how to write history, rather than whether history should be written at all. Every scientific achievement, once available for philosophical reflection, is necessarily a past achievement. Thus some history must be written; the question is "How?"

The history of science written on such a model of scientific rationality will search the chronicles both for scientists who can be construed as advocating the historians' antecedent view of science and for recognized achievements, which the historians themselves must be able to construe as having been achievements on their models of methodology. This is what used to be called a "rational reconstruction." Thus George Sarton, virtual founder of disciplinary history of science in the United States in the early twentieth century, possessed of a positivist definition of science which he explicitly derives from Comte and Mill, declares that science begins "when men began to observe the sun, moon, stars and planets" (Sarton 1962, 1). Sarton conceived science as positive empirical knowledge, the only objective truth attainable, and thus thought of the history of science as "the history of the discovery of objective truth" which proceeds towards that truth by "successive approximations" (Sarton 1962, 102–108).

Sarton's narrative of progress is guided by both methodological criteria and criteria of truth. He includes in his narrative both what he takes to be exemplars of empirical methods and discoveries (or approximations) of what he takes to be timeless truths. And so he praises Leonardo for being able to partially overcome "Platonic and Galenic prejudices" by careful observation and experiment, and Simon Stevin for being the first to represent forces as vectors. If a Leonardo or a Stevin deviated from the positivist way or came to a conclusion which seems false to us, then the "man of science" in them had been somehow perplexed by external forces: for example, Leonardo was "bamboozled by Plato and Galen" regarding the supply of blood in the body (Sarton 1962, 138). More often, however, the perceived deviations from the scientific method and truth are simply not mentioned, and Sarton resorts to listing what he perceives as achievements, presenting these achievements in present terms.

Sarton's criteria are unabashedly those of his present conception of science. And he explicitly uses these criteria to determine whether his historical subjects succeeded or failed in their scientific endeavors. This is not a failure of descriptive history, for Sarton's works are chock full of minute description. Nor is Sarton being naive, for he is very sophisticated at presenting even pre-Socratic science as contemporary. For that reason, it might be helpful to turn to a slightly more transparent example.

Aristotle's *Metaphysics A* is the first example we have of an ahistoricist history of science, and for that reason is very instructive. In this book, Aristotle reviews predecessor theories about the causal composition of nature and shows how, viewed from the standpoint of his completed physics, these predecessors were mistaken and/or only partially successful. Aristotle's own theory was about the causes of phenomena which make them what they both are and appear to be, and employed four types of cause: material (the matter of which a thing is composed), formal (the structure or arrangement of a thing), final (the end or purpose for the sake of which a thing exists), and efficient (the proximate source of motion or change). And so where Thales says that the earth rests on water, Aristotle says that this is evidently a partial description of a material cause. When Empedocles says that Love and Strife are the reason for the mixing and intermingling of earth, air, fire and water, Aristotle says that this is an attempt to specify the efficient cause. According to Aristotle, each of these thinkers begins an intellectual project which culminates in his (Aristotle's) now completed theory. He goes on to say that, if these thinkers knew what he now knows, they would have recognized their error and partiality, and would come to believe in Aristotle's physics.

The first point to notice is the crucial assumption that Aristotle's theory is correct. Otherwise, he could not say that where his predecessors differ with him, they fail; but this is indeed what he wishes to say.

Second, in assuming the correctness of his theory, he employs it as an explicit standard against which rivals are to be judged. Of course, he also appeals to specific phenomena concerning which he claims his theory to have a better account. For instance, he says against Thales that the material principle water will fail to account for locomotion (why should this water move?), whereas the addition of the category of efficient cause could account for locomotion (because the wind drives it). But his appeal to phenomena is itself motivated by a prior aim which is to provide complete causal explanations of natural phenomena, which leads us to a further key point.

Third, and perhaps most important, Aristotle employs his own standards of what science is supposed to be in assessing his predecessors' projects. Thales not only failed to arrive at the correct material cause, he failed to distinguish material from efficient cause. But he must do so if he is to provide a complete causal account of nature, which Aristotle simply assumes Thales is aiming at. This is what makes sense of Aristotle's interpretive move from Thales's assertion that "The earth rests on water" to Aristotle's claim that Thales thought that the material cause of nature is water. Aristotle assumes that Thales is after the same quarry, that Thales's aims are the same as Aristotle's own; and if this is so, then Aristotle must interpret Thales as if he were a contemporary (at least two centuries separate them, along with some large-scale changes in Hellenic culture), engaged in the same project. Thus, "the earth rests on water" must be, on Aristotle's terms, a metaphorical way of asserting an answer to "What is the material cause of nature?"

A physicist I know objected to the previous example by explaining to me that what Aristotle is up to in his physics is either just not scientific, or it is very bad science. For those reasons, she continued, the example is not a good one. But to say this is to employ a tacit criterion of "the scientific" which differs from Aristotle's. This tacit criterion comes from the physicist's own present standards of what constitutes science, and is the source of the evaluations given. And so this objection serves as a confirmation of the function of the tacit standards of the present in ahistoricist history. To read Aristotle either in or out of the history of science requires some conception of the scientific, and the default conception is that employed by the present historian. Thus this contemporary physicist recapitulates Aristotle's historiography by doing to Aristotle what Aristotle did to his predecessors.

To summarize the important points of these anachronistic narratives:

1. Ahistoricist narratives use present scientific aims, truth and methods to either vindicate their subjects by showing how these predecessor theories and practices anticipate the present or vitiate them by showing them mistaken. Not only do they vindicate past scientists who are taken to have approximated present standards, but these narratives, in showing such past scientists to be successful using modern methods, vindicate those present standards themselves. To be sure, the assumption which begins the history is that the standpoint of the present is correct, but that is ex hypothesi: the subsequent narrative vindicates the hypothetical assumption.

2. The standards which they use are those of the present, and a central standard for each is the aim or purpose for which science is done. For Sarton it is positive truth, for Aristotle it is complete and nonaccidental truth. Attributing this aim to predecessors whom they deem "scientists" enables the would-be historian to judge past projects by the same criteria as he would present projects, and neatly arranges the history of science into categories of success, failure, and approximation.

3. The ahistoricist historian must always make an assimilative interpretation of a past author's assertions in order to render them intelligible. Thus, we are told, past authors often speak metaphorically, and the historian's job is to tell us what they literally meant, and this in the vocabulary of the present. As the editors of a recent book on the history of philosophy put it, the historian of philosophy must be a sort of field anthropologist, interpreting in the following way:

> What he said was "The other white god died because he quarreled with the spirit inhabiting the mburi", but what he meant was that Pogson-Smith died because, like an idiot, he'd eaten some of those berries over there. (Rorty et al. 1984, 7)

Two advantages of ahistoricism are that it produces a history which is always intelligible to the present reader and that it gives a clear assessment, not only of who in the past counts as a scientist, but also of the relative merits of past scientists. Intelligibility is virtually assured since the historian produces a narrative which is plausible on the terms of the present. In approaching any figure, the question "What was this scientist trying to do?" is to be answered by assimilating the scientist's aims to those of the present. And so we get a contemporary empiricist reading of Ptolemy in which we find that his sole scientific aim is empirical adequacy, and we find that Bacon, after all, is a hypothetico-deductivist, because he couldn't possibly have seriously believed in inductivism. Historical figures are thus rendered intelligible because they are rendered contemporaries: "pen-friends" across the centuries, as Ian Hacking once put it. Assessment is also easy for the ahistoricist since the standards for achievement are ready to hand, external to the historical scientist under investigation. It is simply a matter of extracting what really counts in a past scientist's work, given what we now know about "the" scientific method and scientific truth.

A final advantage is that it might deepen our understanding of present science by showing the march of progress and error which culminates in contemporary science, displaying the triumph of contemporary methods throughout the ages in reaching the truth despite the obstacles of dogmatism, myth, and religious persecution. This makes the genre for history of science the heroic epic, in which the hero (science) is vindicated in the end by persevering against all odds.

What's wrong with ahistoricism? I think that there are many problems, but perhaps the most poignant is this: in employing contemporary methodological and alethic (of truth) standards, it tends to mistake pervasive features of present science for timeless universal features of "Science." This is, of course, a species of provincialism, a provincialism of the present. But it is a particularly insidious bias, because it ensures that genuine differences in conception, methodology, and truth claims which may be encountered in past scientists will not be appreciated. An extreme version of this is the Sartonesque history which "discovers" that past scientists had exactly the same aims as modern scientists and were possessed with crude forms of the same modern methods. What this history really achieves is only a projection of present standards on the past. But this achievement comes at the expense of falsifying the past and being able to recognize genuine differences in conception and meth-

odology, differences which could prove fruitful alternatives to present conceptions. The failure of methodological consensus among
contemporary philosophers of science points to the need for genuine
alternative methodological conceptions. But if past scientists are to
be read to determine whether they are deductivists, inductivists, or
hypothetico-deductivists, if past methodologies are assimilated to
present categories, then the failures of the present will simply be
projected onto the past. And this precludes the past from being a
source of novelty in the present. The history of science is thus
rendered sterile as a source of novelty, useful only as propaganda.

Perhaps a different approach to history can remedy this problem.

HOW TO WRITE THE HISTORY OF SCIENCE II: HISTORICISM AND THE DENIAL OF GLOBAL STANDARDS

The first full-scale historicist revolution in the West ironically
comes on the eve of the cultural triumph of ahistoricist methodism.
It was a singular insight of Renaissance thinkers that they were able
to suggest that the antiquity which they inherited via Scholasticism
might differ from the antiquity which is recoverable from a careful
study of the ancient works themselves in the original languages and
in their historical context. Of course, humanists like Gassendi did
reject the perceived sterility of Scholastic philosophy, as did Descartes
and Bacon. But unlike these latter methodists, Gassendi and other
humanists did not throw out the baby with the bathwater. Instead,
recognizing that at least some scholastic errors were due to a
corruption and degeneration of the Aristotelian tradition, many
humanists set out to recover what was lost.

The task which Renaissance historiographers set themselves was
the recovery of antiquity on antiquity's own terms. It was a self-
conscious effort to ask what standards and aims ancient authors set
for themselves, and to see those aims, as well as the efforts to attain
them, in a way which was as free as possible from the bias of
Scholasticism. The way to do this was to recognize both the importance of narrative history and the importance of writing a narrative
which was sensitive to a past figure's own aims and standards.
Otherwise, what was recovered by narrative was tainted by corrupt
Scholastic standards.

Gassendi was perhaps more aware than most humanists of the
challenge which Cartesian methodism presented. Lynn Joy (1987)

has argued that Gassendi recognized the ahistoricism of Cartesianism: Cartesianism implies that we can articulate and solve central philosophical and scientific problems independently of their historical situatedness and the intellectual tradition in which they arose. Gassendi responds by analyzing the historical presuppositions of Descartes' methodism. He suggests that we should pay attention to the narrative given by Descartes which places him at an historical crux: this Cartesian history reveals that the tradition of inquiry begun by ancient Greek and Latin figures, notably Aristotle, and handed down in their present day by medieval Scholasticism, is a tradition of myth, bias, error and superstition, and that we would be better off to ignore it entirely by starting over from the beginning. But it is precisely this narrative which allows Descartes to justify his method of demonstration, for his method is rational just because it overcomes and avoids the errors of his predecessors, the errors which have been disclosed by the history he writes; its rationality, at least in part, consists in its historical position. So even if Descartes is right about his methodology, his methodology will be rational only because of its position in history. On Gassendi's view, showing that a scientific method is rational requires that a contextual or historicist narrative be told.

Of course, historicism is not completely vanquished in the period between the sixteenth century and our own, and a careful study of its own history would have to include the eighteenth-century Italian Giambattista Vico, the nineteenth-century Germans Georg Hegel and Friedrich Nietzsche, the nineteenth-century Englishman William Whewell, among others. But perhaps we can pick up the story in the late twentieth century with Thomas Kuhn's attack on the perceived barbarity of ahistoricism. Consider Kuhn's criticism of Sarton's ahistoricism as "the chronicle of the triumph of sound method over careless error and superstition" which yielded "remarkably little information about the content of science" and which not only disguised "both the structure and integrity of past scientific traditions" but constructed a history of sciences "which never existed" (Kuhn 1977, 148–9).

Citing pioneers of this new historicism such as Pierre Duhem, Alexandre Koyré, and Herbert Butterfield, Kuhn says that the writer of historicist narrative should first put aside the contemporary science that he knows. His "science should be learned from the textbooks and journals of the period . . . and he should master these and the indigenous traditions they display." (Kuhn 1977, 110). The historicist should seek to uncover past scientific traditions as coherent wholes; he should "attempt to display the historical integrity of that science in its own time" (Kuhn 1970, 3). In attempting to render the actions and writings of past scientists intelligible, he should ask how those events and writings appeared from the point of view of participants. As Isaiah Berlin put it, the historian must engage in a "kind of imaginative projection of [himself] into the past . . . to capture concepts and categories which differ from those of the investigator" (Berlin in Dray, 1966, 44). Put another way, the historicist should both demarcate the "scientific" and determine success and failure using the aims and standards of the past figures under study.

The history which emerges from this enterprise reveals fundamental discontinuities in the history of "science," discontinuities in conceptions of nature, methodology, and even phenomena. Kuhn's history, like Butterfield's before him, is famous for revealing sharp discontinuities in theory, method, and phenomena. But even Kuhn seems to be conservative about large scale aims in science like coherence, simplicity, and predictive accuracy. He suggests that these cognitive values remain relatively stable through theory change, and provide a continuity to science across revolution, though he admits that the interpretation of these values might be different in different paradigms (Kuhn 1977, 335). However, the accounts of other historicists have suggested that even these general aims change over time. For instance, Ernan McMullin (1988) argues that Aristotle had virtually no interest in predictive accuracy, and would not have recognized it as an aim of science. Robert Westman (1990) argues that Copernicus's own aims were those of a humanist restorer of antiquity, and that the modern traditional view of Copernicus as triumphal empiricist and the later simplicist views distort Copernicus's real achievements because they are conceptually anachronistic.

In what sense does historicist narrative provide justification for science? First, we must recognize that it can only do so for particular scientific traditions, since most historicists agree that there is not sufficient conceptual or methodological continuity to assess groups of traditions by a single set of standards. This is not a "conceptual impossibility" argument to the effect that there can be no cross-traditional standards of methodology and truth. It is, rather, a contingent matter: different traditions of science have in fact employed different standards, and the few standards which might genuinely be cross traditional (say, for example, the inference form we call hypothetical syllogism) both differ in their centrality to different traditions and are insufficient for significant evaluation. Or to put the point a different way, historicists claim that even though there may be universal standards, they are not decisive for all traditions. Moreover, to impose a standard from the outside which is not internal to a scientific tradition, or is internally insignificant (for example, to measure Aristotelian physics by the standard of predictive accuracy), belies a failure to understand what the tradition is up to. Worse, the standard imposed is typically (again a contingent matter) a standard which is both operative within and plausible to present science. A familiar historicist introductory move, in order to show why we need yet another book on some famous figure and do not yet understand him, is to point out that the categories of interpretation and evaluation that other scholarship employs are imported from the present, and so distort our understanding of the real achievements of that figure. For example, Westman argues that other interpretations, by importing contemporary notions of scientific aim and methodology, marginalize Copernicus's political aims and so distort their very large role in his astronomical scheme.

For all these reasons, the historicist narrative constructed in order to establish the success or failure of individuals working within some scientific tradition or other must employ the standards indigenous to that tradition; Newton must be seen to succeed or fail in his own terms, not the terms of the present historian. But it is in precisely this way that such a narrative can be a justification, for it can reveal the triumph of some scientific theory or practice over both its contemporary rivals and the obstacles set for itself. Such a narrative can reveal, for example, Aristotle's teleology as a successful integration of the explanation of not only animate and inanimate objects, but both of those with the explanation of human behavior as well. Judged from the standpoint of modern biology, Aristotle's teleology seems ludicrous; but set within his own tradition, judged on its own terms, it may be justified.

To illustrate, take Westman's Copernican history once more. The problem with ahistoricist views of Copernicus's achievement, Westman argues, is twofold. First, such views must ignore vast amounts of Copernicus's own work and many of his stated aims and heavily weight contextually isolated snippets. For example, the Simplicist view says that one of Copernicus's central achievements was the reduction of the eighty-odd Ptolemaic epicycles to thirty-four, producing far simpler celestial motion. But close attention to Copernicus's technical sections reveals that he more than compensated for this reduction by adding more epicycles and eccentrics when he eliminated Ptolemaic equants. Or when Conceptual Revolutionists say that Copernicus produced a radical intellectual gestalt-shift by seeing the same phenomena in a new way, that view must ignore both his programmatic and technical adherence to circularity and uniform speed, ideals present in, though violated by, both the Ptolemaic and Aristotelian traditions. And all such views ignore the persistent humanist rhetoric and methodology of Copernicus. The discursive practices of humanism pervade Copernicus's writing: quotations from ancient authorities, epistolary prologues, and overt allusions to Horatian aesthetic. Of particular note is Cardinal Schönberg's prefatory epistle approving the movement of the earth, which is routinely (and conveniently) ignored by scholars who prefer to focus on Osiander's unsigned preface, placed in the book without Copernicus's permission. Westman concludes that, taking Copernicus's work as an integrated whole, he is best viewed as a humanist reformer of the clerical order in which he worked, and his specific aim is to reform the astronomical order in order to bring it back into line with more ancient sources (notably Pythagoras and Aristarchos—both heliocentrists), not to revolutionize it.

Second, in imputing aims and standards to Copernicus which he did not himself hold, ahistoricist views render Copernicus a failure. He failed to produce a vastly simpler view than the Ptolemaic tradition, he failed to produce a conceptual revolution, he failed to produce significant new observations. But Westman argues that if we painstakingly reconstruct Copernicus's own aims and standards, in part by reconstructing the clerical humanist tradition of which he took himself to be a part, we discover a humanist reformer of astronomy who by and large succeeds in his limited aims of reordering the universe along aesthetic grounds which derive from Horace's *Ars Poetica*. So Westman's narrative reveals a humanist scientist and cleric who is justified in not abandoning circularity and in not producing a simpler view of the heavens. And curiously, if Westman is right, Galileo will have been the first ahistoricist misreader of Copernicus.

To summarize historicism:

1. Historicist narratives vindicate their subjects by showing them to be successful (when they are, of course) in overcoming their problems and rivals on their own standards of success, problematic and rivalry. Such narratives vindicate or vitiate past science on its own terms. To be sure, some historicists (notably Butterfield) go on to evaluate past standards in light of contemporary ones, but in so doing cease to be historicist.
2. Historicist narrative justification is local. Historicists, in employ-

ing context-dependent standards (that is, standards which are internal to the tradition under study), will only attempt local evaluations of success or failure, and will avoid (as far as possible) assessing past science using the standards of the present. Historicists do not deny the possibility of universal standards. Rather, they typically only assert that none have been found, unless they have been projected from one tradition onto another. Or perhaps the safer historicist claim would be that they can find no universal standards significant enough to demarcate science from non-science or good science from bad.

3. A corollary point is that because the historicist attempts to avoid assimilation, the interpretation of a past author's assertions requires a different sort of anthropological move than that of the ahistoricist. Rather than explaining what past scientists meant on our terms, historicists must say what they meant on their terms, which may sometimes not be intelligible to contemporary readers, if contemporary readers cannot project themselves into the conceptual framework of past tradition.

The central advantage of historicism is that it reveals the integrity and plausibility of past traditions which otherwise may appear ludicrous from the standpoint of the present. For example, Bacon is often ridiculed for being a naive inductivist who thought that particulars (facts) could be collected, ordered, and arranged prior to any theorizing. This can never be, says the proponent of the theory-ladenness of fact, because facts never present themselves pristine. But historicist scholarship (notably Jardine, 1974) reveals a Bacon who was sufficiently an Aristotelian that he assumed all along that topoi (commonplace topics of investigation and argument, deriving from Aristotle's Topics) were available to the investigator, topoi which fit the Aristotelian scheme of contrary predication, such as heat/cold; such topics guided the collection and systematization of particulars. Moreover, Bacon's aim in stressing the collection of all particulars was meant to be a corrective for the narrowness of Scholastic physical topoi, and so is an injunction to expand the range of scientifically relevant topics. An eminently sound strategy if, as Bacon claims, Scholastic science is sterile.

In consequence, a further advantage of historicism is that it reveals past traditions as genuine alternatives to present scientific tradition, rather than as failures or approximations. This revelatory character of historicism is particularly important when present conceptions degenerate and face insoluble problems. Historicist history can be a source of genuine novelty in concept, aim and standard, providing a useful contrast to entrenched and seemingly self-evident ways of thinking.

The historicist project is thus one of recovery, whether Renaissance or contemporary, though of course there are significant differences. But why should this recovery be advantageous to present science? After all, historicists seem to balk at any attempt to adjudicate between past and present traditions of science, or when they do, they merely lapse into ahistoricism and employ present standards, because, of course, we know better now. Indeed, the very idea of adjudication seems somehow out of place, if past traditions are so different. We may compare, but only to highlight the differences.

In fact, the very value of historicism in showing the differences of past scientific traditions seems to preclude the possibility of adjudication. Since, say, the Aristotelian physical tradition differs so greatly from the Newtonian, whose standards should we use to adjudicate? And if they differ on what problems are central, or even on what the problematic is, how can we judge whose problems are better? Historicism seems to give up the ghost of evaluation, in favor of some version of pluralism. This makes the genre of history rather Kafkaesque: isolated portions of the narrative seem coherent, but there is no order or organization to the whole, no progress or decline.

Thus the very large disadvantage of historicism is that, in providing a remedy for the assimilationist tactics of the ahistoricist, it systematically eliminates the normative resources which would enable it to say which tradition is right. Kuhn's point about the lack of neutral observation vocabulary in which to adjudicate disputes between competing paradigms is relatively tame compared to the larger historicist problem of the lack of shared conceptual goals and standards.

Kuhn advocates a fairly mild historicism in which some key elements (theory and phenomenal description) do not remain stable during revolutionary periods, but others (values such as consistency and accuracy) do. And though this was not widely recognized when Kuhn's *Structure of Scientific Revolutions* appeared, this stability during revolutions is the very thing which ensures that paradigm shift is not arational or irrational, not "mob psychology." These are, after all, cognitive values which turn out to be constitutive of scientific rationality. And they are what allow Kuhn to write the continuous narrative of "science" which he does; they are historiographical demarcation criteria. Even if their emphasis and interpretation differ, Kuhn nevertheless thinks that such universal cognitive values are decisive, as other, more radical historicists do not. If it turns out, as Kuhn thinks, that the history of "science" is not all progress, that is a contingent matter—but it is judged progressive or not by transparadigmatic criteria of value.

The problem for a conservative historicist like Kuhn is that even in one of his favorite revolutionary periods, the Copernican revolution, Copernicus turns out not to be so revolutionary, given the value he places on circularity. Moreover, Kepler and Newton overthrow what was a constitutive cognitive value (uniform circularity) for the astronomical tradition from Eudoxos in the sixth century B.C. through Copernicus. Kuhn, of course, admits the first point, but he cannot admit the second, since a value-shift, or rather a shift in the centrality and normative character of a general conception, seriously undermines his stability thesis.

Once more radical historicists than Kuhn assert that there are no significant continuities of normative standard across traditions, then the narrative of the encounters of one tradition with its rivals will be a political narrative, in the pejorative sense of "political." That is, the narrative will be a story of a series of power plays. Of such stuff are the genealogical narratives of Nietzsche and Foucault. Since rationality itself is constituted (on their views) by relationships of power, and since changes in rationality signal power-shifts, genealogy will reveal the dynamic origin of these changes as fundamentally changes of influence, institution, and social order. Rational evaluation of differing traditions is precluded, since rationality is only internal to a tradition, a means of legitimizing its power structure.

What started as a corrective for the barbarity of ahistoricism has seemingly ended in a barbarism of its own, albeit a polar opposite barbarism. But perhaps the opposition is a false dichotomy.

MODERATE HISTORICISM: LOCAL TRUTH AND RATIONAL PROGRESS

Both types of history surveyed so far agree, at least tacitly, that scientific theories and practices are vindicated or vitiated by the narratives that can be told of their successes, modifications, and failures in their encounters with problems and rivals. They differ, however, over the scale on which their narratives can be written. Ahistoricists aspire to a universal history of "science," historicists aspire to a local history of "so-and-so's science." Correspondingly, ahistoricists employ what they take to be universal criteria of rationality and truth in science, the historicist employs local criteria.

Assuming that historicists are right when they detect large-scale differences in scientific traditions, one way to categorize the difference between these two historiographies is to say that ahistoricists employ standards of rational method and truth which are *external* to the past tradition under investigation, while historicists employ standards of rationality and truth which are *internal* to the tradition being examined. Once this distinction is made, we can say that both styles of history are able to reveal a narrative of rational progress and approach to truth, provided we understand that historicists reveal internal rationality and truth and that ahistoricists reveal external rationality and truth. The problem is that on this assumption there turns out to be no external rationality or truth, for the ahistoricists have only revealed their adeptness at provincialism; that is, they have shown how rival traditions fail or succeed on the terms of the present, and in order to accomplish this have produced an interpretive straw man to judge. Remember that our examples of ahistoricism were not ham-handed in proclaiming the correctness of the present; rather, the standards of the present were assumed correct and tacitly employed, their use being warranted by the further assumption that we know more now. Remember too that historicists are always pointing out that the source of "universal" standards of rationality and truth turns out to be present tradition. So we seem to be left with only rival versions of internal rationality and truth in science, and no way to adjudicate between them.

But surely things are not as bleak as they seem. For one thing, I have written as if every tradition succeeds on its own terms. But this is manifestly not the case, as historicist histories have shown. By the late sixteenth century it was commonly recognized, or at least recognizable by those like Galileo, that the medieval Aristotelian tradition in physics had failed to solve outstanding problems and achieve its aims, and had not been able to adequately respond to these challenges with the promise of eventual solution and achievement. And Westman's Copernicus recognizes a degeneration within the Ptolemaic astronomical tradition, a gradual falling away from the standards set by Ptolemy himself. The internal narrative of the failure of the Ptolemaic tradition in astronomy has pretty well been written by now, thanks in large part to Butterfield, Westman, and Kuhn. According to this narrative, the tradition sets itself the problem of calculating apparent planetary movements from two crucial "realities": the geocentrism of the universe and the uniform circular/spherical motion of celestial bodies. As the story unfolds, we see a series of gradually accumulating ad hoc responses to keep the vast theoretical machinery in line with new and more accurate observations: a couple of deferential epicycles here, the addition of an equant there, an adjustment of the orbit's eccentricity, and so on.

Whether we use the language of crisis or degeneration (and for all Lakatos's talk of rational reconstruction, he realizes the crucial role of internal history), nevertheless, the crucial feature of these tradition-internal narratives is that they reveal a history of failure on the tradition's own terms. Thus the genre of internal narrative is not always comic epic; it can be tragedy: the demise of a scientific tradition due to a fundamental flaw in the character of that tradition.

But now a certain rapprochement may be achieved between ahistoricist and historicist narratives. Armed with several internal narratives of ultimate failure, an external narrative can be constructed which reveals those traditions which fail on their own terms. This external narrative may then, using the relatively innocuous external standard "Judge those traditions of science which fail on their own terms to be no longer serious candidates for belief," judge internal failures also to be failures from the standpoint of the present's concern for truth and rationality. It follows that the necessary counterpart to internal narrative is the relatively innocuous external narrative of traditions which fail on their own (internal) terms. For to recognize such internal failure is to recognize that such traditions are no longer serious contenders for our allegiance.

This allows for the sort of evaluative judgments from the standpoint of the present which are necessary for the possibility of justifying present science as the best candidate for truth and rationality so far (not best simpliciter). But this sort of externalist narrative, in contrast to provincial ahistoricism, does not uncritically assume that later is better. Such externalist narratives can be kept honest by first having completed the proper internalist narrative, and then judging the past tradition as it appears in internalist history. The problems of assimilation, straw man, and lack of novelty are addressed while retaining the ability to judge at least those traditions which fail on their own terms. Such failures can still be instructive, but now in a full-blooded fashion, and it may be that key conceptions from failed traditions may be rescued from oblivion to provide genuine alternatives to present conceptions, much as Gassendi's historiography of science can be rescued from its failed attempt to restore an Epicurean atomism.

This provides only limited externalist judgments, however, since it may be that a past scientific tradition is vindicated on its own terms. Aristotelian moral psychology, for instance, may be just such a tradition, perhaps as modified by Thomas Aquinas. Here the attempt at externalist narrative which adjudicates between the Aristotelian tradition and, say, modern developmental theories like Kohlberg's cannot yet get off the ground, since the relevant internal histories have not yet been completed. For to say of a tradition in science that it is vindicated on its own terms is to say that it is a living tradition, a contemporary rival, albeit a rival whose standards differ in radical ways from those of what may be the dominant conception. In such situations, a judgment of which research tradition is the best so far from an external point of view cannot be rendered. We can only pursue whichever tradition we happen to inhabit until internal narrative reveals that one or the other or both have begun to degenerate.

Perhaps the most obvious objection here is that we do seem to evaluate the standards of internally vindicated rival traditions, and this evaluation seems perfectly legitimate. For example, when in the Second Day of Galileo's *Dialogues Concerning the Two Chief World Systems* Salviati (the spokesman for Galileo) says that to attempt physics without geometry is to attempt the impossible, he is offering a pointed criticism of a the Scholastic Aristotelianism of his day. It was widely accepted that geometry and mathematics had little to do with physical science, in large part because Aristotle thought that number was an accidental characteristic of bodies. Galileo is thus attacking the largely nonquantitative standards of (his) contemporary Scholastic physics by proposing precise geometrical proportionality (and measurement) as a sine qua non of physics. But on what grounds is he attacking? It is crucial to point out that Galileo's strategy is to show how his new mathematical science can solve problems of, in this case, the acceleration of heavy bodies, problems which plagued Scholastics. But if this is right, then Galileo's criticism of the nonquantitative character of Scholastic physics is based on a recognition of internal failure to solve problems, problems which, if reformulated in quantitative terms, could be solved. Admittedly, this is only a single (and perhaps fortuitous) example. However, what we learn from this example is that what purports to be an external criticism of a tradition's standards rests upon a prior recognition of the internal failure of that tradition. The specific criticism then amounts to a specific diagnosis of the failure; in this case, Galileo claims that Scholastic physics failed to solve problems of downward acceleration because it lacked sufficiently quantitative standards. All of which is to say that we only seem to attack the standards of a rival tradition without recognizing its internal failure.

A moderate historicism, then, would be a combination of internalist and externalist narrative. The internalist narratives would have to be written first, for the reasons surveyed. Externalist narratives, written from the standpoint of the present, could then be written of those traditions which have failed on their own terms, showing present science to be partially justified precisely because it avoids failed methods and mistaken judgments of truth. Such a narrative would satisfy the historicist aims of preserving the integrity of past science and recognizing genuine differences between traditions while nevertheless adjudicating disputes between them. The crucial stricture remains that no externalist history can yet be written of genuine living, rival traditions in science; one has to wait and see.

HUMANISM REVISITED

Suppose, once again, that Gassendi were the father of modern philosophy, and recognize the force of the counterfactual: that in fact, modern philosophy and the dominant tradition in philosophy of science have been ahistoric*ist* at best, ahistoric*al* at worst. By contrast, I have wanted to suggest both that the humanist emphasis on historical justification has a role to play in philosophy of science and that there are better and worse ways of writing history.

But I promised to answer the question of how contemporary orthodox philosophy of science would differ if the orthodoxy were at least moderately historicist. Here is a brief answer. First, I suspect that the very term "science" would have a different sense and extension than it does, if only for the reason that historicists of science in the latter half of this century have argued that their subject matter is uneven: samples of what count as science in various periods and cultures display no tight cluster of methods, aims or phenomena in virtue of which these samples could be said to be tokens of the same type. Second, temporality would be a central category for philosophy of science. For philosophers would be asking questions like "How did early eighteenth century phlogistic chemistry represent a genuine advance over Boyle's mechanism, even though modern molecular chemistry, sometimes thought to begin with the overthrow of phlogistic theory, has great affinities with Boyle?" instead of "Given the evidence, how can we show that the phlogiston hypothesis has a lower probability than the oxygen hypothesis?" Third, it almost goes without saying that this hypothetical philosophy of science would be at every point historically conscious, aware of the immense complexity and variety which history of science reveals, and loath to engage in contextless debates about the logical relations between "theory T" and "observation O." And last, there would be no need of essays like this one which defended the minimal thesis that history can play a role in the justification of scientific theory and practice.

Of course this essay, though it is an argument about types of historiography, is also itself a sketch of an external narrative about philosophy of science since the sixteenth century. It is a narrative which assumes the internal failure of orthodox ahistorical and ahistoricist philosophy of science, and suggests that humanist historiography, modified in ways which could not have been envisioned by Gassendi and his cohort, nevertheless forms the nucleus of the best available alternative. It was mistake to reject humanist historicist methodology along with its sometimes fawning admiration of antiquity. This is not to say that logic and observation do not have their place. It is rather to say that their place is not foundational nor as central as it has been taken to be, and that narrative justification should be the heir apparent to that place.

Perhaps paradoxically, then, the dominant tradition in philosophy of science, which traces its ancestry to those revolutionaries Descartes and Bacon, is the Scholasticism of the present age. Its problematic has become sterile; it is an impediment to progress. Though this may have been apparent to some for several decades, the recognition is not yet pervasive. Nevertheless we can look forward to the day when future courses in philosophy of science routinely include a couple of weeks on the twentieth century Historiographic Revolution.[1]

---

[1]Portions of this paper were written during the tenure of an NEH Summer Seminar Fellowship. I would like to thank the members of that seminar, particularly Peter Achinstein, for helpful discussion of the issues. Thanks also to Lynn Joy for interpreting Gassendi for me. And as so often, I owe a great debt to Alasdair MacIntyre for thinking these thoughts before me.

REFERENCES

Berlin, Isaiah. 1966. "The Concept of Scientific History" in William Dray, (ed.), *Philosophical Analysis and History.* New York: Harper and Row, 1966, 5–53.

Butterfield, Herbert. 1951. *The Origins of Modern Science, 1300–1800.* New York: Macmillan.

Galilei, Galileo. 1967. *Dialogues Concerning the Two Chief World Systems,* trans. Stillman Drake, 2nd rev. ed. Berkeley: The University of California Press.

Jardine, Lisa. 1974. *Francis Bacon, Discovery and the Art of Discourse.* Cambridge: Cambridge University Press.

Joy, Lynn. 1987. *Gassendi The Atomist: Advocate of History in an Age of Science.* Cambridge: Cambridge University Press.

Kuhn, Thomas. 1957. *The Copernican Revolution.* Cambridge: Harvard University Press.

———. 1970. *The Structure of Scientific Revolutions.* Chicago: University of Chicago Press, 2nd ed.

———. 1977. *The Essential Tension.* Chicago: University of Chicago Press.

Lakatos, Imre. 1978. *The Methodology of Scientific Research Programmes,* vol. 1, John Worrall & Gregory Currie (eds.). Cambridge: Cambridge University Press.

MacIntyre, Alasdair. 1977. "Epistemological Crises, Dramatic Narrative, and the Philosophy of Science," *The Monist 60:* 453–72.

McMullin, Ernan. 1988. "The Shaping of Scientific Rationality" in McMullin, (ed.) *Construction and Constraint.* Indiana: University of Notre Dame Press, 1–48.

Rorty, Richard, J. B. Schneewind, Quentin Skinner, (eds.) 1984. *Philosophy in History.* Cambridge: Cambridge University Press.

George Sarton. 1962. *Sarton on the History of Science,* Dorothy Stimson (ed.) Cambridge: Harvard University Press.

Westman, Robert S. 1990. "Proof, poetics and patronage: Copernicus's preface to *De revolutionibus*" in Westman & David C. Lindberg, (eds.) *Reappraisals of the Scientific Revolution.* Cambridge: Cambridge University Press, 167–205.

INDEX